Innovations in Sustainable Technologies and Computing

Series Editors

Jagdish Chand Bansal, Department of Mathematics, South Asian University, New Delhi, India

Joong Hoon Kim, School of Civil, Environmental and Architectural Engineering, Korea University, Seoul, Korea (Republic of)

Atulya K. Nagar, Liverpool Hope University, Liverpool, UK

AF147681

The book series aims to publish research on the analysis and development of technological innovations that consider natural resources and nurture economic and social development. Innovative sustainable technologies are expected to develop sustainable production planning and tools, which reduces environmental and ecological risks drastically.

The series covers research related to innovative solutions in the field of sustainable technology, computing, and communication. Computational methodologies in the field of computer science and engineering, cybersecurity, data science, information systems and software engineering, algorithms for communication, smart transport system, smart city planning, e-waste management system, and other such sustainable technological solutions within the scope of this series.

The series will publish monographs, edited volumes, textbooks and proceedings of important conferences, symposia and meetings in the field of sustainable technology and computing.

Anil Kumar · Manabendra Saharia

Python for Water and Environment

 Springer

Anil Kumar
Department of Civil Engineering
Indian Institute of Technology Delhi
New Delhi, Delhi, India

Manabendra Saharia
Department of Civil Engineering
Indian Institute of Technology Delhi
New Delhi, Delhi, India

ISSN 2731-880X   ISSN 2731-8818   (electronic)
Innovations in Sustainable Technologies and Computing
ISBN 978-981-99-9410-6   ISBN 978-981-99-9408-3   (eBook)
https://doi.org/10.1007/978-981-99-9408-3

This Springer imprint is published by the registered company Springer Nature Singapore Pte Ltd.
The registered company address is: 152 Beach Road, #21-01/04 Gateway East, Singapore 189721,
Singapore

Paper in this product is recyclable.

To my family,
Late Nilakhi Saharia, Prof. Shrutidhara
Sarma, Ramesh Chandra Saharia, Dhiraj
Saharia, Gunindra Nath Sarma, Pranita
Devi, and others.

—Manabendra Saharia

To my family,
Radhamani Devi, Shukram Puran, Sunil
Kumar, Suneeta Kumari, Anita Kumari,
Sneha Mishra

—Anil Kumar

Foreword by Dr. Martyn Clark

Data science has emerged as a powerful tool to understand changes in the Earth's climate. The use of large amounts of data to produce transformative insights requires a new set of tools and skills. This textbook, "Python for Water and Environment", fills a critical gap between theory, computation, and applications in the field, by enabling readers to implement theoretical concepts and develop an in-depth understanding of water and environment problems.

The objective of the book is not only to serve as a programming guide but also to expose the reader to the unique challenges prevalent in the water and environment sciences. The book uses the versatile Python programming language, which provides a straightforward implementation of models and rapid testing of algorithms. Starting from the basics, the book gradually takes the readers from a basic to an advanced level of programming that is relevant to hydrologic and environmental modeling. The book deals with a wide range of topics such as exploratory data analysis, statistical data modeling, and numerical modeling, all organized into well-defined chapters.

I congratulate the authors for writing this book which can serve as an important resource for researchers and water professionals who wish to include Python in their day-to-day work.

Dr. Martyn Clark
Former Editor-in-Chief (Water Resources Research)
Professor of Hydrology and Schulich Research Chair in Environmental Prediction
Department of Civil Engineering
Schulich School of Engineering
University of Calgary
Calgary, Alberta, Canada

Foreword by Dr. Rangan Banerjee

Water is a critical and scarce resource in India and the world. The challenges of environmental management and sustainable development need new and innovative approaches and new tools and techniques. I have great pleasure in writing the foreword for the textbook "Python for Water and Environment" written by my colleague Manabendra Saharia.

Python has emerged as a language of choice for programmers and researchers. This book demonstrates how Python provides an efficient platform for modeling, analyzing and interpreting data, data analysis, and creating predictive models. This book provides an easy to understand introduction to the use of Python for environmental problems. The author delves into his rich experience in developing models for water and environment problems and presents well-annotated code that can help beginners, practitioners, and experts.

I hope this textbook helps propel the next generation of researchers and future professionals to develop software and analytical tools and techniques for solving the problems of water and the environment.

We at IIT Delhi hope that this textbook will serve as a bridge between theory and practice and serve to propel new interdisciplinary research in this important area. The author will be happy to get your feedback and suggestions on further enhancements in this book and the domain.

Dr. Rangan Banerjee
Director
Indian Institute of Technology Delhi
Forbes Marshall Chair Professor in the Department of Energy Science
and Engineering (IIT Bombay)
Fellow of the Indian National Academy of Engineering

Preface

While teaching graduate students of Civil Engineering at IIT Delhi, we felt the need for a textbook that focuses more on the practical implementation side of Water Resources Engineering using a modern programming language, which could supplement the excellent theoretical textbooks that already exist. In an era where data science and machine learning have revolutionized all fields, students continue to struggle with breaking into specialized domains that require increasingly advanced computational skills. Thus, this book focuses on code examples that readers can directly benefit from. By providing concrete examples, the book equips readers with the skills needed to address the complex challenges faced by water and environmental professionals in today's rapidly changing world.

"Python for Water and Environment" is conceived as a practical guide for professionals, researchers, and students who are working in sectors of water and environment. This preface outlines our journey through the realm of Python programming, where we venture into the science of water resources and environmental management. The essence of this book lies in the seamless integration of theoretical principles with computational prowess, harnessing the power of Python to model, analyze, and solve real-world problems.

Our aim is to illuminate the potential of Python as a robust and versatile tool for dealing with complex challenges in the domain of water and environment. We aim to break down the barriers to entry that have traditionally existed for non-programmers. We believe that being an open-source language Python aligns well with the shared global responsibility of water and environmental management. This book, therefore, goes beyond being a mere guide to Python programming. Whether you are a seasoned professional or a passionate beginner in this domain, "Python for Water and Environment" is designed to be your companion in this exciting journey of discovery and problem-solving.

July 2023   Dr. Manabendra Saharia
Assistant Professor
Indian Institute of Technology Delhi
New Delhi, India

Dr. Anil Kumar
Principal Project Scientist
Indian Institute of Technology Delhi
New Delhi, India

Acknowledgements

We extend our heartfelt gratitude to the Indian Institute of Technology Delhi community, for providing us with an intellectually challenging academic environment. The Department of Civil Engineering deserves a special mention for its continual support and encouragement, enabling us to explore and expand the horizons of our professional expertise.

Manabendra Saharia dedicates this book to his mother, Late Mrs. Nilakhi Saharia, whose life of hardwork and kindness he aspires to live up to. He would also like to acknowledge the support of his loving wife (Prof. Shrutidhara Sarma), father (Ramesh Chandra Saharia), brother (Dhiraj Saharia), father-in-law (Gunindra Nath Sarma), and mother-in-law (Pranita Devi). He acknowledges the extraordinary debt he owes to all his well-wishers over the years: Prof. Rajib Bhattacharjya, Prof. Sharad K. Jain, Prof. Parthajit Roy, Prof. Parthasarathi Choudhury, Prof. G. V. Ramana, Prof. Sumedha Chakma, Prof. D. R. Kaushal, Prof. B. R. Chahar, Prof. Pierre Kirstetter, Dr. Jonathan J. Gourley, Prof. Yang Hong, Dr. Sujay Kumar, Dr. Augusto Getirana, Dr. Andy Wood, Dr. Andy Newman, Prof. Martyn Clark, and many more. He also acknowledges the friendships that have sustained him over the years.

Anil Kumar would like to dedicate this book to his mother (Mrs. Radhamani Devi), father (Shukram Puran), brother (Sunil Kumar), and sisters (Suneeta Dhungia and Anita Kumari). He is grateful to his well-wishers: Prof. Kumar Hemant Singh, Prof. Mohan Yellishetty, Prof. Trilok Nath Singh, and Prof. Stuart D. C. Walsh. He acknowledges his friendship with Dr. Sneha Mishra and Dr. Rohit Kumar Shrivastava for their constant support.

We specially acknowledge the encouragement and leadership of the Director of the institute, Prof. Rangan Banerjee, and the Head of the Department of Civil Engineering. Prof. Arvind K. Nema. Without their steadfast support, this book wouldn't see the light of day. Finally, we would also like to express our sincere thanks to the many reviewers (Dr. Aatish Anshuman, Prof. B. R. Chahar, Ms. Reetumoni, etc.) who took the time to meticulously scrutinize our work. Their invaluable insights and constructive feedback played an instrumental role in shaping this book. They challenged us to refine our ideas, improve our methodology, and ensure standards of quality and accuracy in writing.

Contents

About the Authors

Dr. Anil Kumar is a senior project scientist in the Department of Civil Engineering at the Indian Institute of Technology Delhi. He received his Ph.D. in Computational Geosciences jointly from Monash University (Australia) and the Indian Institute of Technology Bombay (India). He received a B.Tech. in Geophysical Technology from the Indian Institute of Technology Roorkee. He has been working as a researcher in the field of machine learning and numerical modeling and has helped develop innovative solutions for the oil, gas, and mining industry.

Dr. Manabendra Saharia is an assistant professor in the Department of Civil Engineering and an associate faculty member of the Yardi School of Artificial Intelligence at the Indian Institute of Technology Delhi. Prior to joining IIT Delhi, he held positions in the hydrology labs of the NASA Goddard Space Flight Center and the National Center for Atmospheric Research (NCAR). He received his Ph.D. in Water Resources Engineering from the University of Oklahoma. At IIT Delhi, his HydroSense research lab focuses on utilizing physics and data-driven techniques to monitor and mitigate natural hazards such as floods and landslides. He has been recognized for his scientific contributions, having received Young Scientist awards from both the National Academy of Sciences, India (NASI), and the International Society for Energy, Environment and Sustainability (ISEES).

Part I
Practical Python for a Water and Environment Professional

Chapter 1
Data Analysis in the Water and Environment

1.1 Introduction

Water security is increasingly in jeopardy throughout the world. Too much water causing floods, too little water causing droughts, or poor water quality affecting health can endanger life, economy, and ecosystems. In order to detect, monitor, and mitigate these diverse problems in water and environment, we require actionable intelligence based on data. Data analysis is an important part of understanding the complex and multidimensional relationships between water systems and their surrounding environments. We investigate these relationships through a combination of science, engineering, and technology, which will help in discovering new information relevant to the impact of water and the environment on human life. The efficacy of our strategies in managing water resources, preserving aquatic life, combating pollution, and preparing for climate change rests on our ability to monitor, model, and mitigate various types of water hazards.

Water sustains life and is one of our planet's most precious resources. It acts as an integral link in the vast chain of ecosystems that allows life to prosper. Various environmental factors such as geological formations and anthropogenic activities impact the quality, availability, and distribution of water. These exchanges result in a perpetually evolving ecosystem, making it necessary to constantly monitor this water-environment interface.

Data has been dubbed as the new oil. And just like crude oil, raw data has to be refined and analyzed to extract valuable and meaningful insights. Understanding patterns, trends, and relationships in data can help us in making inferences about the state and performance of water and environment systems, their interactions, and the potential effects of changes in one system on the other. Data analysis helps us assess the impacts of an oil spill on coastal waters, study seasonal variations in river flow, or model future scenarios of sea-level rise due to global warming.

Data analysis in the water and environment sector involves a combination of methodologies and technologies, ranging from traditional statistical techniques to advanced machine learning algorithms. It relies on data acquired from a variety

© The Author(s), under exclusive license to Springer Nature Singapore Pte Ltd. 2024   3
A. Kumar and M. Saharia, *Python for Water and Environment*, Innovations in Sustainable Technologies and Computing, https://doi.org/10.1007/978-981-99-9408-3_1

of sources, such as satellite images, weather stations, sensor networks, and socio-economic databases, to explore and interpret complex phenomena related to water and the environment. By leveraging computational power and algorithms, vast amounts of data can be processed and analyzed, resulting in actionable insights for scientists, policymakers, and stakeholders. Data analysis helps us understand the impact of human activities such as industrial pollution, deforestation, overfishing, and uncontrolled urbanization on water and the environment. Such insights help in the formulation of better policies and effective strategies for sustainable development, water management, and environmental conservation. Data analysis helps in the optimization of resources and predictive modeling for future scenarios. Data analysis thus enables us to anticipate and mitigate risks, harness opportunities for sustainable growth, and create a better balance between human needs and environmental preservation.

However, data analysis in the water and environment sector is not without its challenges. The quality and integrity of data, the complexity of environmental systems, the inherent uncertainty in many types of environmental data, and the need for multidisciplinary approaches are among the many issues that analysts must grapple with. These challenges necessitate a continuous refinement of methods and techniques, emphasizing the field's dynamic and evolving nature.

Data analysis in the water and environment sector integrates scientific exploration, use of technology, and consideration of natural phenomena. The process involves diving into challenging datasets, and uncovering details of environmental processes. It is a journey of exploration, problem-solving, and creating impact, which is crucial in an era marked by environmental shifts, climate change, depleting resources, and rising human needs. In this chapter, we shall investigate the techniques, significance, and function of data analysis in the water and environment domain.

1.2   Types of Data

Hydrologists study the distribution, movement, and storage of water in the environment, which requires large amounts of data across diverse scientific and engineering disciplines. The data collected in hydrology is generally of three types: spatial, temporal, and attribute data.

Spatial data consists of the geographic and physical properties of an area of interest. In hydrology, examples of spatial data are soil and rock type, topography, the physical features of water bodies, and vegetation. These attributes influence of movement of water within the water cycle. For example, how water moves across the land surface is influenced by topographical attributes such as elevation, slope, and aspect. Geographic Information System (GIS) tools are widely used for managing and analyzing spatial data, which can be collected using different means such as satellite remote sensing, LiDAR, or ground observations.

When numerical data is tracked over time, it is called temporal data. Extracting valuable insights about the trends, patterns, and shifts in temporal attributes is known as temporal data mining. In hydrology, examples of temporal data include temperature, snowpack thickness, groundwater levels, station precipitation, and evapotranspiration. Ground-based measurements, weather stations, and satellites can collect these hydrological measurements. Time series data are crucial for identifying annual and seasonal patterns, understanding the impact of climate change, and predicting hydrological scenarios in the future.

Attribute data, on the other hand, is any qualitative description that provides additional context regarding the spatial and temporal data. For example, it may involve data on population density and socioeconomic indicators in the study area, which can provide insights into human impacts on the hydrological system. Typically, attribute data is collected using field measurements, laboratory analysis, public databases, and direct surveys.

These categories of data often overlap and lead to rich and multilayered datasets which can be explored from different perspectives. For instance, water quality parameters (attribute data) may be measured over time (temporal data), and across a variety of locations (spatial data) within a watershed. Such a dataset can be explored to understand how water quality in an area changes spatially as well as temporally. Increasingly, in hydrology, the problems have become complex enough that it requires harnessing big data and machine learning to develop deeper insights into large, complex, and multidimensional datasets. By integrating data from diverse sources, hydrologists can unravel the complex dynamics of water in the environment, and aid in the formulation of effective water management strategies and sustainable practices.

In the later chapters, we delve into hydrological data modeling using Python. We explore time series analysis and numerical modeling of water variables such as streamflow, water level, and atmospheric pressure. We will show how Python's powerful libraries can be leveraged for stochastic and deterministic modeling and trend detection. We will also demonstrate the use of Python for solving partial differential equations in the context of groundwater flow, and solute transport modeling. These steps and techniques will equip the reader with the tools necessary for the analysis of complex water systems.

Chapter 2
Python Environment and Basics

2.1 Integrated Development Environment (IDE)

Integrated Development Environment (IDE) and virtual environments are two crucial components of Python programming. They are key tools in a Python developer's toolkit, aiding in streamlining the development process, enhancing productivity, and ensuring code reliability and reproducibility.

An IDE is a software application that provides comprehensive facilities to programmers for software development. For Python developers, using an IDE like PyCharm, Jupyter Notebook, or Visual Studio Code has numerous benefits. One of the key advantages is that it combines several tools and features needed for coding into a single interface. This includes text editors for writing and editing code, debuggers for finding and fixing errors, syntax highlighting for better readability, and auto-completion features that save time by suggesting completions for names of functions, keywords, and variables.

Furthermore, many IDEs provide built-in support for version control systems like Git, allowing developers to track changes, revert to previous versions of code, and efficiently collaborate with other developers. IDEs are built with functions for code refactoring, testing, and profiling, which are essential for writing clean, error-free, and efficient code.

2.2 Why Virtual Environments?

Virtual Environments are isolated environments where python packages can be installed without interfering with other projects and installations. This allows the developer to create isolated environments for different projects, which can be communicated to other developers who can recreate the exact development environment for those projects. Without virtual environments, installing different versions of the

A. Kumar and M. Saharia, *Python for Water and Environment*, Innovations in Sustainable Technologies and Computing, https://doi.org/10.1007/978-981-99-9408-3_2

same package for different projects can lead to conflicts and inconsistencies, making it difficult to share or deploy code. Virtual environments solve this problem by creating isolated spaces for each project, where the necessary dependencies can be installed without the risk of interference.

Virtual environments also contribute to the reproducibility of Python code. Maintaining a record of the exact versions of the packages used in a project, allows other programmers and systems to recreate the environment and execute the code under identical conditions. This is particularly important in scientific computation and data analytics, where package dependency is a basic requirement.

Both the Integrated Development Environments (IDEs) and virtual environments play crucial roles in Python programming. While IDEs increase efficiency and code quality by integrating utilities and features into a unified interface, virtual environments ensure the reliability and replicability of Python code by ensuring project dependencies and isolation. Together, they simplify the coding process by providing a versatile framework for Python programming and making it more enjoyable.

2.3   The Anaconda Package Manager

The anaconda package manager (https://www.anaconda.com/) provides an open-source distribution of the Python and R programming languages for scientific computing, designed for simplified package management and deployment. It is loaded with a collection of over 1,500 packages for computation, data analysis, visualization, and machine learning, and incorporates fundamental libraries such as NumPy, Pandas, and Matplotlib. In addition, it is equipped with a package manager, Conda, that assists in managing environments, making the installation and maintenance of different versions of software packages hassle-free.

> **macOS Installation:** To install Anaconda on your operating system, start by downloading the Anaconda installer for your operating system from the official Anaconda website. Python 3.8 version or later is the preferred choice unless a specific earlier version is required. Upon downloading, double-click the downloaded file to launch the installer. This program guides the user through the installation process, possibly suggesting the installation of the Anaconda3 version in the home user directory. After completing the steps, the installation is verified by opening a new terminal window and typing the command: "conda list". If Anaconda is installed correctly, a list of installed packages appears.
>
> Anaconda's popularity is due to the convenience it offers. Operating systems, like MacOS and Windows Subsystem for Linux, come preinstalled with Python, which might not be suitable for scientific computing due to its potential incompatibility with certain modules. Thus, having a separate, controlled environment like Anaconda ensures compatibility between Python and other packages.
>
> **Linux Installation:** The installation process for Anaconda on a Linux system begins with downloading the installer script from the Anaconda website. After

downloading, the user needs to verify the data integrity of the installer with cryptographic hash verification through the SHA-256 checksum. A terminal is opened and the bash command is run on the downloaded file, like this: "bash Anaconda3-2020.02-Linux-x86_64.sh". "2020.02" would be the version downloaded. Read the license agreement by scrolling with the help of the "Enter" key and accept it by typing "yes". Next, choose an installation location or accept the default location. Once the installation process is complete, the terminal is closed and a new one is opened to ensure the changes take effect.

Anaconda's ability to handle dependencies and environments on Linux systems provides a significant advantage as Linux distributions comprise many interdependent packages. Anaconda offers a reliable and easy-to-use interface to manage the complexity of these packages.

Windows Installation: On Windows, the installation process begins by downloading the Anaconda installer .exe file from the official Anaconda website. Once the installer is downloaded, it is run and the instructions are followed. During installation, the user will be asked if they want to add Anaconda to the PATH environment variable—it is recommended not to check this option, as it can interfere with other Python installations or software. Rather, one can use Anaconda software by launching Anaconda Navigator or the Anaconda Command Prompt from the Start Menu.

Windows lacks preinstalled Python, so installing Anaconda is the simplest way to get started with Python, especially for beginners. It can seamlessly handle complex Windows environments, making it effortless to manage and distribute a large number of third-party libraries which frequently pose difficulties during installation and maintenance in the Windows ecosystem.

2.4 The Jupyter Notebook

The Jupyter Notebook is an open-source web-based application that has emerged as a preferred tool among data scientists, academicians, and researchers worldwide. Its simple but powerful user interface allows users to produce and share documents enriched with live code, mathematical equations, data visualizations, and explanatory text.

One of the key features of a Jupyter Notebook is its flexibility in accommodating more than 40 programming languages, including Python, R, Julia, and Scala making it highly adaptable to diverse computational requirements and projects. Users can toggle between these languages inside the notebook, making it an exceptional medium for polyglot data analysis.

Jupyter Notebooks can embed multimedia elements like images, videos, LaTeX scripts, and JavaScript widgets, making it easy to write narrative documents interspersed with code and results. This feature is also suitable for crafting presentations, dashboards, and interactive tutorials. It also allows connections to multiple data sources, allowing users to pull data from databases, APIs, or cloud storage.

For those already proficient in Python or other compatible programming languages, gaining expertise in Jupyter Notebook is simple. The user interface is quite beginner-friendly, helping users swiftly learn how to execute cells, tweak code, and visualize data. A wide array of tutorials and resources is available online to assist newcomers.

A notable feature of a Jupyter Notebook is that it allows executing code in "cells". This enables users to efficiently segment their work into manageable units, beneficial for code debugging and revision. Each cell operates independently, allowing for iterative code development, testing, and validation of algorithms and functions.

The coding style in Jupyter Notebook fosters an investigative and iterative coding methodology. As it encourages the concept of "literate programming", users can include with their code, notes ,and comments, leading to improved legibility and sustainability. This style is apt for structured and lucid data analysis workflows.

Jupyter Notebook's compatibility with Git and GitHub also extends its accessibility. Users can set version control for their notebooks, share them among peers, and even publish them right away as web pages via GitHub.

To summarize, the Jupyter Notebook's versatility, user-centric design, and support for collaborative and investigative programming make it an invaluable tool for a diverse user base, from researchers involved in complex data analyses to educators instructing in programming or data science disciplines.

2.5 Installing External Packages

Installing external packages in an Anaconda virtual environment is a process that involves several steps, ensuring that the necessary software tools are correctly added to the user's workspace. The Conda package and environment manager included in Anaconda are used to handle these installations.

1. **Creation of a new Conda environment:** Before installing external packages, it is good practice to create a new Conda environment. This isolated environment provides a separate space where packages can be installed without interfering with the system's base Python installation or other Conda environments. This can be executed using the command:

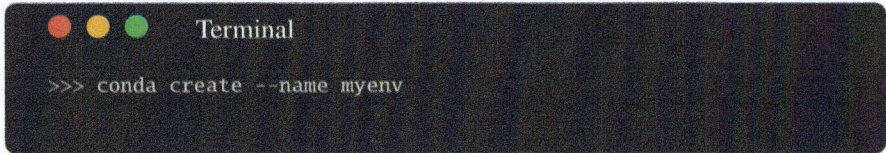

```
Terminal
>>> conda create --name myenv
```

A new name of the environment can be provided in place of "myenv". As per requirement, a Python version may be specified by appending "python=x.x" (where x.x is the version number) to the command.

2. **Activating the new environment:** Prior to installing any packages, the environment must be activated. This can be done with the command:

```
Terminal
>>> conda create --name myenv
>>> conda activate myenv
```

3. **Installing packages:** Depending upon the source, the packages can be installed after activating the environment, either by using the "conda" or "pip" command. It is typical to use Conda when the package exists in the Anaconda distribution, given its effectiveness in handling dependencies. Issuing the following command initiates the installation:

```
Terminal
>>> conda create --name myenv
>>> conda activate myenv
>>> conda install package-name
```

The name of the required package can replace the "package-name". The pip command can be used in the case when the package is not available in the Anaconda distribution. Pip can access packages uploaded to the Python Package Index (PyPi). To use the following command, do: "pip install package-name", again, replacing "package-name" with the name of the needed package.

4. **Verifying the installation:** It is a good idea to verify the installed packages. Issuing the "conda list" or "pip list" commands displays the installed packages in the active environment:

```
● ● ●          Terminal
>>> conda create --name myenv
>>> conda activate myenv
>>> conda install package-name
>>> conda list
```

The aforementioned steps permit users to install external packages in a Conda environment, which are ready to be stored and shared, making it easy to duplicate the development configuration across varied and distributed groups.

Chapter 3
Python Essentials

3.1 Getting Started with Python

Python is the most popular programming language in the data analysis world and the growing significance of data has made mastering Python an important skill to acquire. Due to its simplicity and code readability, Python is a popular choice among beginners and professionals alike. A wide set of libraries makes it a perfect tool for tasks spanning from data analysis and machine learning to visualization and web development.

Getting started with Python requires setting up an appropriate environment. Beginners may prefer Anaconda, a platform that conveniently bundles Python with many data science libraries. Online resources, such as Python's official documentation, offer a wealth of knowledge for self-learners. For a structured learning experience, several online courses are available on platforms like Coursera and edX. Practical coding exercises on websites like HackerRank and Codecademy can also be useful. Python's readability and extensive community support make the learning journey manageable and rewarding. It goes without saying that persistent practice along with practical projects is crucial to mastering Python.

3.2 Setting Up Python Environment

Preparing a Python environment calls for installing Python, selecting an integrated development environment (IDE), and overseeing package management. Python can be installed via the Anaconda manager or from its official website. PyCharm or Jupyter Notebook are feature-rich IDEs that offer robust functionalities for software development. Package management utilities, such as pip and conda, assist in managing dependencies across libraries. Python virtual environments provide project-level

© The Author(s), under exclusive license to Springer Nature Singapore Pte Ltd. 2024   13
A. Kumar and M. Saharia, *Python for Water and Environment*, Innovations in Sustainable Technologies and Computing, https://doi.org/10.1007/978-981-99-9408-3_3

isolation to maintain specific dependencies, ensuring error-free operation. The reader is referred to Sects. 2.3, 2.4, and 2.5 for setting up the Python environment.

3.3 My First Python Script

Python is a powerful and flexible programming language which is appealing for beginners due to its readability and straightforwardness. Here's an example of a simple Python code that asks for the name of the user and then welcomes them:

```python
# This is a simple Python script

# Ask the user for their name
name = input("What's your name? ")

# Print a greeting to the user
print(f"Hello, {name}! Welcome to Python programming.")
```

```
● ● ●        Terminal

>>> What's your name? >? Roy
>>> Hello, Roy█ Welcome to Python programming.
```

This script employs the "input()" function for in-taking user input, to store in the variable termed, "name". The "print()" function then displays a salutation that incorporates the user's name.

The "f" before the string enables string formatting, which allows the embedding of variable's value directly within the string object. The "{}" brackets denote insertion point where the variable's value will be inserted when the "print()" statement is called.

Python annotations start with the "#" character. Python does not execute subsequent content after "#" thus serving as a convenient way to add comments or explain the functioning of the code.

This script shows some of the basic aspects of Python such as inclusion of variables, mechanisms for user input, print statements, and string formatting schemes. This is an introductory code for beginners, paving way to explore more elaborate concepts such as loops, conditional statements, function definitions, and data structures.

3.4 Python Fundamentals

3.4.1 Basic Syntax

This sample code demonstrates "single" and "multi" line annotations. It also illustrates how to define a function and display the returned result.

```python
1   # This is a single line comment in Python
2
3   """
4   This is a
5   multi-line comment (or docstring)
6   in Python
7   """
8
9   # Import the built-in math module
10  import math
11
12
13  # Define a function
14  def calculate_circle_area(radius):
15      """
16      This function calculates the area of a circle
17      given a radius.
18      """
19      if radius < 0:
20          print("Error: Radius cannot be negative.")
21          return
22      area = math.pi * (radius ** 2)
23      return area
24
25
26  # Call the function
27  radius = 10
28  area = calculate_circle_area(radius)
29  print(f"The area of the circle with radius "
30        f"{radius} is {area:.2f}")
```

```
●  ●  ●        Terminal

>>> The area of the circle with radius 10 is 314.16
```

The program starts by importing the built-in math module. It defines a function named calculate_circle_area(), which takes a radius as a parameter and calculates the area of a circle using the formula, $Area = \pi r^2$. The function checks if the radius is negative and prints an error message if so. The $**$ operator is used for exponentiation in Python.

The function is then called with a specific radius, and the resulting area is printed using Python's formatted string literals, also known as f-strings (f"..."). The 2f inside the f-string is used to format the area result to two decimal places.

This small program illustrates various aspects of Python syntax including comments, importing of modules, defining functions, conditional statements, arithmetic operations, function calls, and formatted print statements.

3.4.2   Functions

In Python, a function is a named block of reusable code designed to perform a particular task. Functions are defined using the "def" keyword followed by the function name and parentheses "()". Inside these parentheses, zero or more parameters can be specified. The function body is indented and typically includes a "return" statement. Functions promote code reusability, improve readability, and allow modular programming.

```python
def outer_function(outer_arg):
    print(f"Outer function argument: {outer_arg}")

    # This is a nested or inner function
    def inner_function(inner_arg):
        return outer_arg * inner_arg

    # Calling the nested function
    return inner_function(5)


result = outer_function(10)
print(f"Result from inner function: {result}")
```

```
● ● ●          Terminal
>>> Outer function argument: 10
>>> Result from inner function: 50
```

In the above program, the outer_function() is defined with a single parameter, outer_arg. Inside the function, it prints the value of the outer_arg using f-string formatting. Within the outer_function(), there is a nested or inner function called inner_function(). This inner function takes a parameter inner_arg. The inner_function() multiplies the outer_arg with the inner_arg and returns the result. After defining the nested function, the outer_function() calls the inner_function() with the argument 5. The return value of the inner_function() is then stored in the variable result.

The program then prints the value of the result using f-string formatting.

3.4.3 List and Tuples

Both lists and tuples are sequence types that can store multiple items. Lists are mutable, meaning you can modify their content, making them great for data manipulation. They are defined using square brackets. Conversely, tuples are immutable and are defined using parentheses, often used for fixed data.

```python
# List Creation
fruits_list = ["Apple", "Banana", "Cherry",
               "Dates", "Elderberry"]

# Accessing List Items
print("First fruit in the list: ", fruits_list[0])

# Modifying List Items
fruits_list[1] = "Blueberry"
print("Fruit list after modification: ", fruits_list)

# Adding Items to the List
fruits_list.append("Fig")
print("Fruit list after adding a new item: ", fruits_list)

# Removing an item from the list
fruits_list.remove("Cherry")
print("Fruit list after removing an item: ", fruits_list)


# Tuple Creation
fruits_tuple = ("Grapes", "Honeydew", "Ice-Apple")

# Accessing Tuple Items
print("First fruit in the tuple: ", fruits_tuple[0])

# Trying to modify tuple items (This will result in an error)
# fruits_tuple[1] = "Jackfruit"

# Tuples are immutable, which means you can't add or
# remove items after the tuple is defined.

# Thus, we can't demonstrate adding or removing items
# in a tuple like we did with the list
```

```
●  ●  ●        Terminal
>>> First fruit in the list:  Apple
>>> Fruit list after modification:  ['Apple', 'Blueberry',
↵   'Cherry', 'Dates', 'Elderberry']
>>> Fruit list after adding a new item:  ['Apple', 'Blueberry',
↵   'Cherry', 'Dates', 'Elderberry', 'Fig']
>>> Fruit list after removing an item:  ['Apple', 'Blueberry',
↵   'Dates', 'Elderberry', 'Fig']
>>> First fruit in the tuple:  Grapes
```

In the above program, first, a list named "fruits_list" is created with five initial elements. The code then showcases accessing list items by printing the first fruit in the list using indexing. Next, an item in the list is modified by assigning a new value to the second element. The updated list is printed to show the modification. After that, a new item, "Fig," is appended to the list using the append() method. The list is printed again to display the addition of the new item. Finally, an item, "Cherry," is removed from the list using the remove() method, and the resulting list is printed.

In the second part, a tuple named "fruits_tuple" is created with three initial elements. The code demonstrates accessing tuple items by printing the first fruit in the tuple using indexing. It also attempts to modify a tuple item by assigning a new value to the second element, but this operation generates an error. This highlights that tuples are immutable, meaning their items cannot be modified after creation.

3.4.4   Dictionaries and Dataframes

A dictionary in Python is an unordered collection of data values. It is also used to store data values like a map, which enables users to hold key:value pairs. Unlike other data types, it holds only a single value as an element.

A dataframe is a two-dimensional labeled data structure with columns of potentially different types. It is generally the most commonly used pandas object.

```
1    # Importing pandas module
2    import pandas as pd
3
4    # Define a dictionary containing river details
5    river_details = {'Name': ['Ganges',
6                              'Brahmaputra',
7                              'Yamuna',
8                              'Godavari',
9                              'Narmada'],
10                'Length (km)': [2525,
11                                3848,
12                                1376,
13                                1465,
14                                1312],
15                'Drainage Area (km2)': [1080000,
16                                        651334,
17                                        366223,
18                                        312812,
19                                        131200]
20                    }
21
22   # Convert the dictionary into DataFrame
23   river_df = pd.DataFrame(river_details)
24
25   # Print the data frame
26   print(river_df)
```

```
●  ●  ●          Terminal
>>>             Name  Length (km)  Drainage Area (km2)
>>> 0         Ganges         2525              1080000
>>> 1    Brahmaputra         3848               651334
>>> 2         Yamuna         1376               366223
>>> 3       Godavari         1465               312812
>>> 4        Narmada         1312               131200
```

The given Python program stores some details about Indian rivers in a dictionary and then converts that dictionary to a dataframe using pandas.

The output of this script will be a dataframe that contains the names of the rivers, their lengths in kilometers, and their drainage areas in square kilometers. Each river represents a different row in the dataframe, and the details (name, length, and drain kilometers) represent the columns.

3.4.5 Loops

Given is a simple Python program that illustrates the use of both "for" and "while" loops. This program prints the first ten numbers (1–10) using both types of loops.

```python
# Using a for loop
print("Using a for loop:")
for i in range(1, 11):
    print(i)

# Using a while loop
print("\nUsing a while loop:")
i = 1
while i <= 10:
    print(i)
    i += 1
```

```
●  ●  ●        Terminal

>>> Using a for loop:
>>> 1
>>> 2
>>> 3
>>> 4
>>> 5
>>> 6
>>> 7
>>> 8
>>> 9
>>> 10
>>> Using a while loop:
>>> 1
>>> 2
>>> 3
>>> 4
>>> 5
>>> 6
>>> 7
>>> 8
>>> 9
>>> 10
```

In this code example, the "for" loop along with the "range()" function acts as a number generator that generates a series of numbers from 1 to 10. Within the loop, the "print()" function is used to display the numeral in the console. The "while" loop runs unconditionally until the condition "i" is less than or equal to 10 returns a true value. Again, the "print()" function is used to print the current value of "i" as the loop iterates through the sequence. When the value of "i" is more than 10, the condition "i <= 10" no longer returns true, consequently terminating the loop.

3.4.6 Conditional Statements in Python

Here, we give a Python program that uses conditional statements to find out whether an input number is positive, negative, or zero.

```python
# Take input from the user
num = float(input("Enter a number: "))

# Use conditional statements to check the number
if num > 0:
    print("The number is positive")
elif num == 0:
    print("The number is zero")
else:
    print("The number is negative")
```

```
●  ●  ●          Terminal
>>> Enter a number: >? 23
>>> The number is positive
```

In the given program the "if" statement checks if the number is greater than zero. If this condition holds a true value, it displays "The number is positive". If the criterion returns a false value, it goes to the "elif" statement. The "elif" (an abbreviation for "else if") statement checks whether the input number is zero. If this condition returns true, it displays "The number is zero". If this condition is also invalidated, it goes to the else statement, which takes care of all other criteria not covered by the preceding "if" and "elif" statements. That essentially means the input number must be less than zero, printing the message "The number is negative".

3.4.7 File Operations in Python

Here, we show a simple Python code that illustrates some basic file operations such as writing to a file, reading from a file, and appending to a file.

```python
1   # Writing to a file
2   with open('myfile.txt', 'w') as file:
3       file.write('Hello World!\n')
4       file.write('This is a simple Python script '
5                  'illustrating file operations.\n')
6
7   # Reading from a file
8   with open('myfile.txt', 'r') as file:
9       print("Contents of myfile.txt:")
10      print(file.read())
11
12  # Appending to a file
13  with open('myfile.txt', 'a') as file:
14      file.write('This line was appended to the '
15                 'file.\n')
16
17  # Reading from a file after appending
18  with open('myfile.txt', 'r') as file:
19      print("Contents of myfile.txt after "
20            "appending:")
21      print(file.read())
```

```
●  ●  ●        Terminal

>>> Contents of myfile.txt:
>>> Hello World!
>>> This is a simple Python script illustrating file operations.
>>> Contents of myfile.txt after appending:
>>> Hello World!
>>> This is a simple Python script illustrating file operations.
>>> This line was appended to the file.
```

Upon initiation, the code writes two lines to "myfile.txt". It then reads the contents of the file and displays them. A new line to the file is subsequently written, and the contents are read again and printed in the console, revealing that line.

Within the active directory, it's crucial that permissions for read/write be given. To avoid accidental data overwriting, one should exercise caution when writing a file.

Chapter 4
Exploratory Analysis of Hydrological Data

4.1 Examining a Dataset

The first step in any investigation is to assess and delve into the dataset. It lets you understand data attributes, spot possible anomalies, and formulate well-grounded choices concerning the statistical techniques suitable for the following analysis. For instance, the nature of data—continuous or discrete, Gaussian distributed or skewed, or manifesting cyclic patterns—would dictate the selection of analytical methodologies. Comprehending the temporal and spatial data resolutions is crucial in hydrology studies as it can influence the interpretation of the results. Preliminary inspection of the data is also helpful in spotting errors, outliers, and missing data points that may affect the study's validity. Thus, by getting well-acquainted with their dataset before undertaking a more comprehensive analysis, learners can work towards a robust and dependable result, thereby reducing the chances of false interpretations or erroneous conclusions. So, examining a dataset is vital for deriving robust findings in hydrologic studies.

4.1.1 Types of Data

Hydrology refers to the scientific study of the distribution and movement of water. They describe a variety of data types to model, analyze, and predict water-related phenomena in the hydrological cycle. These data types can be broadly classified into spatial data, temporal data, and data from hydrological model outputs.

1. **Spatial Data:** Spatial data are data that describe or contain information about objects, events, and phenomena related to Earth's surface. These include topography, land use, soil type, vegetation cover, etc. For example, a Digital Elevation Model (DEM) provides topographic data, which is essential for understanding catchment characteristics, flow direction, and flow accumulation in hydrological

© The Author(s), under exclusive license to Springer Nature Singapore Pte Ltd. 2024   23
A. Kumar and M. Saharia, *Python for Water and Environment*, Innovations in Sustainable Technologies and Computing, https://doi.org/10.1007/978-981-99-9408-3_4

modeling. Land use and soil type data inform about infiltration rates, evapotranspiration, and runoff generation. Vegetation data can help determine transpiration rates. These data are generally represented as raster or vector formats in Geographic Information Systems (GIS) for hydrological analysis and modeling.

2. **Temporal Data:** Temporal data are observations that change over time. In hydrology, these typically involve time series data of precipitation, temperature, wind speed, humidity, river discharge, groundwater levels, and so on. Rainfall data, for example, are valuable for anticipating floods and managing water resources. Data pertaining to temperature and humidity can contribute to evapotranspiration rates, while river discharge data are critical for understanding river dynamics, evaluating flood risk, and engineering hydraulic constructs. Depending on the various use cases, temporal data can be logged at various intervals—hourly, daily, monthly, and yearly.

3. **Hydrological Model Outputs:** These constitute data generated from computational hydrological models that simulate different facets of the water cycle, spanning from rainfall-runoff dynamics and evapotranspiration to soil water migration, and river patterns. The models employ spatial and temporal data inputs and yield outputs like streamflow, runoff, soil moisture indices, and groundwater levels. These data are often used to learn about interlinked hydrological processes, anticipate future water-related occurrences, and make wise decisions about water management.

While these data types are commonly used, it is important to recognize that hydrology is a multifaceted, interdisciplinary field. It frequently requires combining data from various other domains such as meteorology, geology, and ecology. This presents a host of opportunities for developing novel approaches for data analysis, data modeling, and interpretation. Given the importance of water in sustaining life and biotic systems, the study and understanding of these data types are paramount for tackling various water-related challenges facing the world today.

Depending on the number of variables they contain, datasets can also be categorized into univariate, bivariate, or multivariate. Each of these categories provides distinct insights and analytical approaches.

1. **Univariate datasets in Hydrology:** Univariate datasets in hydrology comprise of observations of a single variable over time or space. Typical examples are daily rainfall measurements at a specific location, measurement of river discharge over time at a certain point along a river, or soil moisture levels measured at varying depths at a particular site. Univariate analysis enables the quantification of the central tendency, spread, and distribution of the recorded variable, and is frequently employed to identify trends, patterns, and anomalies within a single variable.

2. **Bivariate datasets in Hydrology:** Bivariate datasets consist of observations of two distinct variables. It is often helpful in understanding whether two variables are correlated and what is the nature of this relationship. An example could be contrasting the amount of rainfall with river flow rates to understand the variation of runoff with precipitation. Bivariate analysis can employ techniques

from regression analysis to quantify the strength and nature of the relationship between the two variables.

3. **Multivariate datasets in Hydrology:** Multivariate datasets consist of observations of more than two variables. In hydrology, such datasets are very common, as water processes are influenced by numerous factors. For instance, a dataset might include precipitation, temperature, evaporation, runoff, and soil moisture, all recorded over time at a particular location. Multivariate analysis allows for the exploration of interactions and relationships among multiple variables simultaneously. This can be crucial in understanding complex hydrological processes and systems. Techniques used can include multiple regression, factor analysis, and cluster analysis, among others.

The choice of univariate, bivariate, or multivariate analysis in hydrology largely depends on the research question at hand and the available data. Regardless of the approach, each provides valuable insights into the characteristics and behavior of hydrological variables, thus contributing to our understanding and management of water resources.

4.1.2 Basic Data Characteristics

Data is the backbone of hydrological studies, and its diverse nature requires careful examination and understanding. The characteristics of the data significantly impact the way it is analyzed, interpreted, and visualized. Below are some of the key data characteristics in the context of hydrology:

1. **Level of Measurement:** Data in hydrology can span all four levels of measurement. For instance, nominal data might represent different land use categories or types of soil in a catchment. Ordinal data could represent the pollution level of water bodies, categorized as low, medium, or high. Interval data could include temperature measurements, which, depending on the scale used, may lack a true zero point. Finally, ratio data commonly encountered in hydrology include measurements like rainfall depth, river discharge, and groundwater level, all of which have a true zero point and consistent scale.
2. **Discrete vs. Continuous:** Discrete data in hydrology might include the number of rainy days in a month or the number of flood events in a year. Continuous data are extensively found in hydrology—rainfall intensity, river flow rate, and water table depth, to name a few.
3. **Univariate, Bivariate, and Multivariate:** The complexity of hydrological processes often necessitates dealing with multivariate data. However, univariate data analysis, such as the study of a single variable like rainfall or temperature over time, is common. Bivariate data analysis looks at the relationship between two variables, rainfall and runoff, while multivariate data analysis could involve multiple variables like temperature, precipitation, evaporation, and soil moisture.

4. **Temporal and Spatial:** Hydrological data often have both spatial and temporal components. For example, rainfall data could be collected at multiple locations (spatial) over time (temporal). Understanding these components is essential for predicting patterns and making decisions about water resource management.
5. **Missing Values:** Missing values are a common issue in hydrological data due to reasons like sensor malfunctions, data entry errors, or inaccessible measurement locations during extreme weather conditions. Techniques such as interpolation or imputation may be used to handle these missing values.
6. **Outliers:** Outliers, while sometimes considered anomalies or errors, can also represent extreme but legitimate hydrological events such as floods or droughts. Identifying and correctly handling outliers is crucial to accurate data analysis and data modeling.
7. **Skewness and Kurtosis:** Many hydrological datasets, such as rainfall and river flow data, do not follow a normal distribution. They might be skewed (asymmetrical) or have high kurtosis (heavy-tailed). Understanding these characteristics can guide the selection of suitable statistical models.
8. **Dependency:** Many variables in hydrology are dependent. For instance, runoff depends on factors like rainfall, soil type, and land cover. Recognizing these dependencies is important for building accurate hydrological models.

Understanding the basic characteristics of hydrological data is the key to performing appropriate and meaningful data analysis. It helps in the selection of suitable statistical techniques, the interpretation of results, and ultimately, the generation of reliable insights into hydrological phenomena. Whether one is developing a predictive model for flood events, studying the impacts of climate change on water resources, or planning for sustainable water management, a solid grasp of these data characteristics is essential.

4.1.3   Common Variable Types

Understanding variable types is crucial for a researcher or a data analyst as the type of variable often dictates the methods of analysis, visualization, and interpretation. The variable types can be broadly classified into four categories: nominal, ordinal, interval, and ratio.

1. **Nominal Variables:** Also known as categorical variables, nominal variables categorize data without implying any sort of order or hierarchy. Examples of nominal variables include the color of a car, the breed of a dog, or the type of soil in a specific location. Each category is unique and one isn't inherently better or worse than another. In data analysis, nominal variables are often used to segment data into distinct groups for comparison.
2. **Ordinal Variables:** Like nominal variables, ordinal variables also categorize data, but they introduce the concept of order. Ordinal variables represent a ranking or ordering of data, but the distances between ranks may not be equal. An example

might be a customer satisfaction survey where respondents can rate a product as "poor", "fair", "good", or "excellent". In this case, "excellent" is clearly better than "good", but we don't know by how much.

3. **Interval Variables:** Interval variables are numerical variables for which the intervals between values are consistent and meaningful, but there is no true zero point. Temperature, for instance, is an interval variable because the difference between 20 and 30° is the same as the difference between 70 and 80°. However, there is no absolute zero point because you can have negative temperatures.
4. **Ratio Variables:** Ratio variables, like interval variables, have a consistent, meaningful scale, but they also have a true zero point. This allows us to say that one value is twice as large as another. Examples of ratio variables include height, weight, and age.

Understanding the type of each variable in a dataset is crucial as it indicates the appropriate statistical methods to be used. For instance, while you can calculate the mean for interval and ratio data, it doesn't make sense to calculate the mean for nominal or ordinal data. Similarly, while you can perform a t-test on interval or ratio data, you would use a chi-square test for nominal data. Therefore, accurately identifying variable types is essential in data analysis.

4.2 Summarizing a Dataset

Summarizing a dataset involves understanding its key statistical attributes. Probability distributions depict the spread and likelihood of data points. Central tendency metrics, like the mean, median, and mode, reveal the "center" of the data. Variability measures, such as standard deviation or variance, illustrate data dispersion around the central tendency. Skewness represents symmetry and kurtosis represents "tailedness", indicating data asymmetry. Gaussianity, pivotal in many statistical tests and models, checks whether the data aligns with a normal distribution.

4.2.1 Theoretical Probability Distributions and Applications

Theoretical probabilistic distributions are the fundamental tools in statistics, data analysis, and machine learning. They describe the likelihood of outcomes in a set of data, allowing us to make inferences and predictions about future observations. There are numerous theoretical distributions, each with distinct characteristics, that make them suitable for different types of data and applications.

One of the most well known is the Gaussian or Normal distribution. It is widely applied in natural and social sciences due to its mathematical properties and the Central Limit Theorem. For example, it is used in quality control to identify products or services deviating from set standards. In finance, it forms the basis for many models, such as portfolio optimization and option pricing, despite its limitations in capturing extreme events.

Binomial distribution describes the number of successes in a fixed number of independent Bernoulli trials, each with the same probability of success. It is widely used in quality assurance, election polling, and risk assessment, among others.

Poisson distribution, on the other hand, models the number of events happening at a fixed interval of time or space, given a constant mean rate of occurrence. It is applied in fields like telecommunications, insurance, and traffic management, where we model events such as the number of calls received by a call center, the number of claims in an insurance company, or the number of cars passing through a toll booth.

Exponential distribution is used for modeling the time between events in a Poisson process, commonly used in reliability studies and survival analysis. For example, it can be used to model the time between failures of a machine or the survival time of patients after treatment.

Theoretical probabilistic distributions also play a critical role in machine learning algorithms, such as the Gaussian mixture models used in unsupervised learning, or the Naive Bayes classifier, which leverages conditional probabilities based on Bayes' theorem.

Hydrology, the study of the distribution and movement of water in the environment, often involves probabilistic analysis of different variables. Below are some of the commonly used probability distributions in hydrology:

1. **Normal Distribution (Gaussian Distribution):** This distribution is used when the data is symmetric and bell-shaped, which often is the case with many hydrological variables. For instance, it is used in analyzing annual precipitation, river flow, and other general meteorological variables.

   The probability distribution function (PDF) is given by

   $$f(x|\mu, \sigma^2) = \frac{1}{\sqrt{2\pi\sigma^2}} \exp -\frac{(x - \mu)^2}{2\sigma^2} \tag{4.1}$$

   The cumulative distribution function (CDF) is given by

   $$F(x|\mu, \sigma^2) = \frac{1}{2}\left[1 + \mathrm{erf}\left(\frac{x - \mu}{\sigma\sqrt{2}}\right)\right] \tag{4.2}$$

   where

   - μ is the expectation or the mean of the data.
   - σ is the standard deviation of data.
   - σ^2 is the variance of the data.

2. **Log-Normal Distribution:** Log-normal distribution is frequently employed to model variables in hydrology that exhibit skewness, such as sediment loads, river flows, or groundwater levels.

   The probability distribution function (PDF) is given by

$$f(x|\mu, \sigma^2) = \frac{1}{x\sqrt{2\pi\sigma^2}} \exp -\frac{(\log x - \mu)^2}{2\sigma^2} \tag{4.3}$$

The cumulative distribution function (CDF) is given by

$$F(x|\mu, \sigma^2) = \frac{1}{2} + \frac{1}{2}\mathrm{erf}\left[\frac{\log(x) - \mu}{\sqrt{2}\sigma}\right] \tag{4.4}$$

where

- μ is the mean of the logarithm of the data.
- σ is the standard deviation of the logarithm of the data.

3. **Exponential Distribution:** It is common to use this distribution to model the time between events, such as the time between rainstorms or the time until the failure of a hydraulic structure.
   The probability distribution function (PDF) of the distribution is given by

$$f(x|\lambda) = \lambda \exp \lambda x \tag{4.5}$$

The cumulative distribution function (CDF) of the distribution is given by

$$F(x|\lambda) = 1 - \exp \lambda x \tag{4.6}$$

where λ is called the rate parameter and is equal to the reciprocal of the mean of the distribution.

4. **Gumbel Distribution:** The Gumbel distribution is also known as the Extreme Value Type I distribution. In hydrology, it is used to model extreme events such as annual flood maxima and extreme rainfall events. The probability distribution function (PDF) of the distribution is given by

$$f(x|\mu, \beta) = \frac{1}{\beta} \exp -\frac{(x - \mu)}{\beta} \exp -\exp \frac{-(x - \mu)}{\beta} \tag{4.7}$$

The cumulative distribution function (CDF) of the distribution is given by

$$F(x|\mu, \beta) = \exp -\exp \frac{-(x - \mu)}{\beta} \tag{4.8}$$

where

- μ is the mode of the distribution.
- β is the scale parameter.

4.2.2 Summarizing Numerical Data

In any data analysis, first, we explore the most common statistical features of a dataset such as central tendency, variability, and symmetry. The mean, median, and mode represent measures of central tendency and provide a quick snapshot of what could be considered as the "average" or "typical" value of the dataset. Measures of variability such as the range, interquartile range, variance, standard deviation, and coefficient of variation are also crucial. They indicate how data points are spread around the center and their deviation from the average. Symmetry measures such as skewness and kurtosis jointly characterize the shape of the data distribution. They help us to understand how data scatters around the mean, informing whether the data distribution is peaked or flat and whether it is symmetric or skewed.

Collectively, these measures provide a comprehensive statistical summary and insight into the underlying distribution and characteristics of the dataset. In hydrology, such statistical summaries are essential for understanding and predicting water-related phenomena and events.

4.2.2.1 Measures of Central Tendency

In hydrology, measures of central tendency are vital to grasp patterns of water-related variables, such as precipitation, groundwater levels, and streamflow. The mean, mode, and median are the measures used for determining the dataset's central tendency, each providing valuable insight and helping in understanding the typical values to be expected.

1. **Mean:** The mean, also known as the arithmetic mean, is calculated by dividing the total sum of all data points by the total count of data points. Every data point is accounted for in this central tendency measure. The formula for calculating the mean, (μ), is

$$\mu = \frac{1}{N} \sum_{i=1}^{N} x_i \tag{4.9}$$

   where N is the total number of observations, and (x_i) is each individual observation.

   For example, if we have monthly rainfall data for a year (in mm): 80, 90, 100, 110, 120, 130, 100, 110, 120, 130, 140, 150, the mean rainfall is (80+90+100+110+ 120+130+100+110+120+130+140+150)/12 = 114.17 mm.

2. **Median:** The median is the middle value in a dataset ordered from the smallest to the greatest. If the number of observations (N) is odd, the median is the value at position ($N + 1$)/2. If N is even, the median is the average of the values at positions $N/2$ and ($N/2$) + 1.

For example, if we consider the same rainfall dataset and order it: 80, 90, 100, 100, 110, 110, 120, 120, 130, 130, 140, 150, the median rainfall is the average of the 6th and 7th observations (110 and 120), which is 115 mm.

3. **Mode:** The mode is the value(s) that appears most frequently in a dataset. A set of data may have one mode (unimodal), more than one mode (multimodal), or no mode at all (no clear peak frequency).
   Continuing with the rainfall dataset, the data points 100, 110, 120, and 130 all appear twice, making them all modes of the dataset. Therefore, the dataset is multimodal.

The measures of central tendency provide a simple and effective way to summarize and understand data. However, they should be used in conjunction with other statistical measures (such as dispersion measures) to generate a complete picture of the data distribution. For example, two datasets could have the same mean but very different spreads.

In hydrology, these measures of central tendency are vital for analyzing and predicting water-related events. Mean values indicate typical scenarios, whereas the mode and median offer insights into repeating trends and exceptional occurrences. Grasping the central tendency of hydrological data proves valuable in planning and managing water resources, forecasting floods, and addressing droughts.

4.2.2.2 Measures of Variability

Variability in data, sometimes also known as dispersion, is an essential aspect of statistical analyses as it measures the extent or spread of data points within a dataset. Primary variability indicators include the range, interquartile range, variance, standard deviation, and coefficient of variation. These measures are crucial in hydrological studies, as they provide insights into the fluctuations and uncertainties associated with different water-related variables like rainfall, river flow rates, or soil moisture content.

1. **Range:** The range is a simple measure of variability. One can calculate it as the difference between the highest and lowest values in the dataset. Consider the example of the annual river flow rates (in cubic meters per second) for a particular location over ten years: 50, 55, 60, 65, 70, 75, 80, 85, 90, and 95, the range would be 95–50 = 45 m^3/s.
2. **Interquartile Range (IQR):** The interquartile range, or IQR, signifies the central 50% of a dataset. It is the difference between the upper quartile (Q3) and the lower quartile (Q1). Q1 is the median of the first half of the data, while Q3 is the median of the second half. In the above river flow rates data, Q1 is the median of the first five observations (50, 55, 60, 65, 70) which is 60 m^3/s, and Q3 is the median of the last five observations (75, 80, 85, 90, 95) which is 85 m^3/s. Therefore, IQR = Q3 – Q1 = 85 – 60 = 25 m^3/s.
3. **Variance:** Variance measures how far each number in the dataset is from the mean (μ) and thus from every other number in the set. It is often denoted by σ^2. The formula for variance is

$$\sigma^2 = \frac{1}{N-1} \sum_{i=1}^{N} (x_i - \mu)^2 \tag{4.10}$$

For the river flow rates data, we first, calculate the mean (μ) which is 72.5 m³/s, then use the formula to get the variance.

4. **Standard Deviation:** Standard deviation (σ) is the square root of the variance. It is useful because, unlike the variance, it is in the same units as the data. Thus, it provides a measure of variability that can be easily interpreted in the context of the data. Using the variance calculated previously, the standard deviation is simply its square root.

5. **Coefficient of Variation (CV):** The coefficient of variation represents the ratio of the standard deviation to the mean, and it is often expressed as a percentage. It is useful for comparing the degree of variation from one data series to another, even if the means are drastically different from each other. The formula for CV is

$$CV = \frac{\sigma}{\mu} \times 100\% \tag{4.11}$$

Again, using the mean and standard deviation calculated earlier, one can easily find the CV.

In hydrology, measures of variability are critical for analyzing the distribution and the fluctuation in various datasets, which can aid in the design and management of hydraulic structures and water resource systems. These measures also help quantify the uncertainties associated with hydrological predictions and estimations.

4.2.2.3 Measures of Symmetry

Symmetry in statistical analysis describes the shape and spread of the distribution of data. Skewness and kurtosis are the two common measures of symmetry that indicate the nature of the distribution of a dataset. Both skewness and kurtosis can provide important insights into hydrological variables like rainfall distribution, river flows, groundwater levels, etc.

1. **Skewness:** Skewness measures the extent and direction of asymmetry in a data distribution. A symmetrical dataset will have a skewness of 0. If the dataset is skewed to the left (or negatively skewed), it means the left tail of the distribution is longer and the majority of the data are concentrated to the right. If the dataset is skewed to the right (or positively skewed), it means the right tail of the distribution is longer and the majority of the data are concentrated to the left. The formula for skewness (γ) is

$$\gamma = \frac{1}{N} \sum_{i=1}^{N} \left(\frac{x_i - \mu}{\sigma} \right)^3 \tag{4.12}$$

where N is the total number of observations, x_i is each individual observation, μ is the mean, and σ is the standard deviation. For example, let us say we have a dataset of annual rainfall (in mm) of a specific area: 100, 110, 120, 130, 150, 200, 250. This dataset is positively skewed, as it has a long right tail, indicating that there is a majority of the less rainfall events, with a few high rainfall events.

2. **Kurtosis:** Kurtosis measures the "tailedness" of a data distribution. A distribution with high kurtosis exhibits a distinct peak near the mean that declines rapidly and has heavy tails. On the other hand, low kurtosis distributions have a flat peak around the mean that declines slowly and has light tails. The formula for kurtosis κ is

$$\kappa = \frac{1}{N} \sum_{i=1}^{N} \left(\frac{x_i - \mu}{\sigma} \right)^4 - 3 \tag{4.13}$$

It must be noted that "-3" is often added to the calculation to provide a comparison to the normal distribution, which has a kurtosis of 3.

For example, a dataset of river flow rates during a flood event would typically display high kurtosis, with most flow rates being relatively stable and a few extreme flow rates representing the peak of the flood.

In hydrology, skewness can provide insights into the tendency of the occurrence of certain events. For instance, positively skewed rainfall data may indicate the occurrence of a few heavy rainfall events. Kurtosis can help hydrologists understand the probability of extreme events. A higher kurtosis in a river flow dataset could indicate a greater chance of extreme flood events. By understanding the symmetry of the distribution, hydrologists can better analyze and predict the behavior of various hydrological variables (Figs. 4.1 and 4.2).

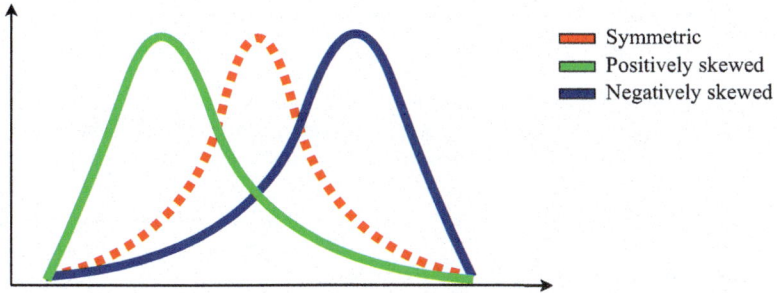

Fig. 4.1 Example of symmetric, positively, and negatively skewed distribution

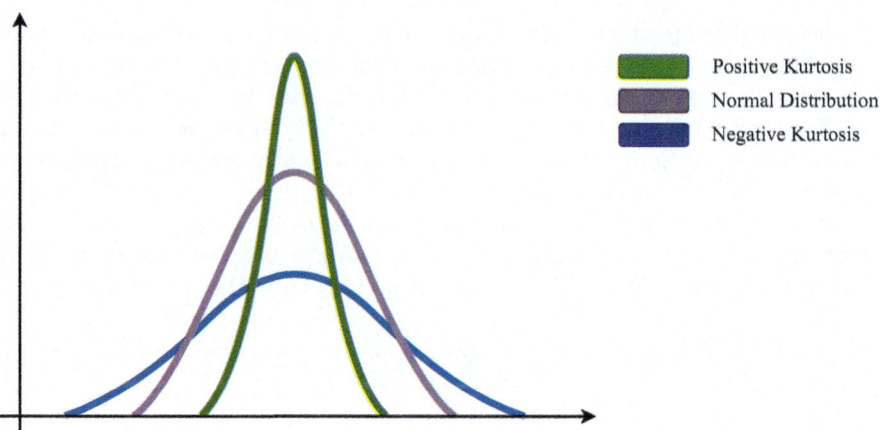

Fig. 4.2 Example of a Normal distribution in comparison with a distribution that has positive kurtosis and a negative kurtosis

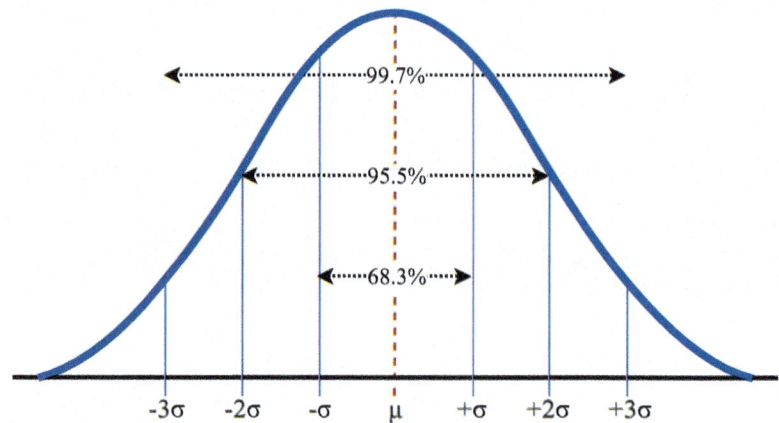

Fig. 4.3 Illustration of a normal distribution where the mean, median, and mode coincide. It also shows the number of standard deviations from the center. Within the first standard deviation, approximately 68.3% of data falls, 95.5% fall within two standard deviations, and 99.7% fall within three standard deviations

4.2.3 Gaussianity in Numerical Data

Gaussianity refers to the tendency of a dataset to follow a Gaussian distribution, also known as a normal distribution or bell curve. This distribution is characterized by its symmetrical bell-shaped curve where the mean, median, and mode coincide at the center of the distribution (Fig. 4.3).

4.2.3.1 Advantages of Gaussianity

1. **Simplicity and universality:** The Gaussian distribution has well-defined statistical properties and can be described by just two parameters: the mean and the variance. Many natural phenomena and processes, including those in social, physical, and biological sciences, tend to follow a Gaussian distribution due to the central limit theorem.
2. **Predictive analysis:** The Gaussian distribution is particularly useful for predictive analysis because its well-defined statistical properties can simplify the estimation of probabilities and quantiles. This is particularly valuable in risk assessment and management.
3. **Parametric statistical tests:** Many statistical tests and procedures (like t-tests, ANOVA, and linear regression) are based on the assumption of data being normally distributed. These tests are powerful and efficient when their assumptions are met.
4. **Data transformation:** Data can often be transformed to achieve Gaussianity, which can help in the application of appropriate statistical methods. Common transformations include the logarithmic, square root, and Box-Cox transformations.

4.2.3.2 Disadvantages of Gaussianity

1. **Misinterpretation:** The simplicity of the Gaussian distribution can lead to misinterpretation or over-simplification of data. Real-world data can be messy and complex, with skewness, kurtosis, multiple peaks, or heavy tails that a Gaussian distribution cannot capture.
2. **Non-applicability:** Not all data follow a Gaussian distribution. For example, incomes, populations, or website visitors often follow a power law or log-normal distribution. In such cases, applying Gaussian-based analysis can lead to incorrect conclusions.
3. **Sensitive to Outliers:** The Gaussian distribution is sensitive to outliers. A small number of extreme values can significantly skew the mean and inflate the standard deviation, distorting the representation of the data.
4. **Violation of Assumptions:** Many parametric statistical tests assume Gaussianity, but real-world data often violate this assumption. Applying these tests to non-Gaussian data can lead to biased or misleading results.
5. **Independence Assumption:** Gaussian distribution assumes that all data points are independent, which is not the case in many datasets, like time series data or spatial data, where values are often correlated with their neighbors.

In conclusion, Gaussianity is an important property that simplifies data analysis and provides valuable insights when its assumptions are held true. However, when dealing with non-Gaussian data, it is crucial to be aware of its limitations and to consider alternative distributions or non-parametric methods. The main goal of data analysis is to understand and accurately represent the underlying reality, where the objective guides the choice of distribution.

4.2.4   Limitations of Summary Statistics

For developing deep insights about a data, it is helpful to know the summary statistics such as the central tendency, measures of variability, and measures of symmetry. However, it is also important to acknowledge their limitations during the analysis. Here, we list some of the reasons:

1. **Sensitivity to outliers:** The mean, as a measure of central tendency, is sensitive to extreme values or outliers. A single outlier can significantly skew the mean, providing a misleading representation of the dataset's central point. Variance and standard deviation, as measures of variability, are also affected by outliers as they account for the squared differences from the mean.
2. **Ignorance of data distribution:** While measures like median and mode are less sensitive to outliers, they may not always provide a complete picture of the data. For instance, the mode is not useful for continuous data or data with no repeating values. Moreover, the mean, median, and mode do not provide any information on the actual distribution of the data.
3. **Loss of detail:** Summary statistics inherently involve a reduction in data, leading to a loss of specific details. The range, for instance, only considers the maximum and minimum values, disregarding how data is spread between these extremes.
4. **Misleading interpretation of skewness and kurtosis:** While skewness and kurtosis can inform about data asymmetry and "tailedness", interpreting them can be challenging. For example, a dataset with skewness close to zero could be symmetric, or it could be a bimodal distribution with two equally sized modes at opposite ends.
5. **Not suitable for complex multivariate data:** Summary statistics mainly provide a snapshot of univariate data. When dealing with multivariate data or when relationships among data are crucial, these measures may not suffice. Correlation, regression, or more advanced techniques might be needed.

In summary, while these summary statistics offer a useful first step in understanding a dataset, they should not be used in isolation. A comprehensive analysis of data should consider these limitations and employ graphical tools or advanced statistical techniques to supplement these measures. The goal should always be to capture as much information about the underlying distribution and relationships in the data as possible.

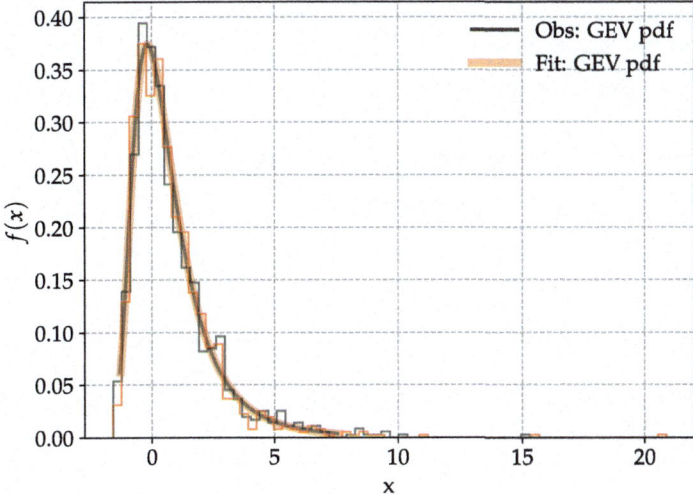

Fig. 4.4 Maximum likelihood estimate (MLE) of the parameter of the Generalized extreme value distribution (GEV) using the Downhill simplex algorithm. The thin black and red lines show the histograms corresponding to the observed data with true parameter (c) and synthetic data from estimated parameter (c_{est}), respectively

4.2.5 Fitting a Distribution

Maximum Likelihood Estimation (MLE) is an effective statistical method used for estimating the parameters of a probability distribution or statistical model. It is founded on the principle of likelihood, which measures the compatibility of the observed data with specific parameter values. The MLE technique aims to find the values of parameters that maximize the likelihood function, making the observed data more probable. Simply put, it attempts to estimate the parameters that are "most likely" to have produced the observed data.

One key advantage of MLE is its consistency. As the sample size increases, the MLE converges in probability to the true parameter value. It is also efficient in achieving the lowest possible variance among all consistent estimators.

However, MLE can be sensitive to initial assumptions about the distribution of data. It may also produce biased estimates for small sample sizes, a problem often addressed by using bias-corrected MLE. Next, we describe a small example of maximum likelihood estimation.

```python
"""
Program to do a maximum likelihood
estimate (MLE) with generalised
extreme value (GEV) distribution
"""

from scipy.stats import genextreme
from scipy.optimize import fmin
import numpy as np
import matplotlib.pyplot as plt


# Fix random seed
np.random.seed(21)


# Likelihood function (the GEV distribution)
def likelihood_fun(theta, data):
    lik = -np.sum(
        np.log(
            genextreme.pdf(x=data, c=theta)
        )
    )
    return lik


# Loss function (to be maximised)
def fit_distribution(data):
    x0 = np.random.rand(1, )
    estimated_params = fmin(
        func=likelihood_fun,
        x0=x0,
        disp=True,
        args=(data,))
    return estimated_params


# Generate synthetic data
c = -0.2  # True parameter for GEV distribution
```

```
40  rvs = genextreme.rvs(c, size=1000)   # Random variates
41
42  # Fit the parameters
43  c_est = fit_distribution(data=rvs)   # Estimated parameter
44
45  # Print results
46  print('Obs. c = {:.4f}'.format(c))
47  print('Est. c = {:.4f}'.format(c_est[0]))
48
49  # Analysis
50  x_obs = np.linspace(genextreme.ppf(0.01, c),
51                      genextreme.ppf(0.99, c), 100)
52  x_est = np.linspace(genextreme.ppf(0.01, c_est),
53                      genextreme.ppf(0.99, c_est), 100)
54  rvs_ = genextreme.rvs(c_est, size=1000)
55
56  # Display the estimate
57  fig, ax = plt.subplots()
58  ax.plot(x_obs, genextreme.pdf(x_obs, c),
59          'k-', lw=2, alpha=0.9, label='Obs: GEV pdf')
60  ax.plot(x_est, genextreme.pdf(x_est, c_est),
61          'r-', lw=5, alpha=0.4, label='Fit: GEV pdf')
62  ax.hist(rvs, color='k', density=True, bins='auto',
63          histtype='step', alpha=0.7)
64  ax.hist(rvs_, color='r', density=True, bins='auto',
65          histtype='step', alpha=0.7)
66  ax.set_xlabel('x')
67  ax.set_ylabel('$f(x)$')
68  ax.grid(ls='--')
69  ax.legend(loc='best', frameon=False)
70  plt.show()
71  plt.tight_layout()
72  plt.savefig('MLE_fit.pdf', dpi=300)
```

The above code generates synthetic data and performs maximum likelihood estimation (MLE) using the generalized extreme value (GEV) distribution.

We first import the required libraries: "scipy.stats.genextreme()" for the GEV distribution, "scipy.optimize.fmin()" for optimization, "numpy" for numerical operations, and "matplotlib.pyplot" for plotting.

Next, the likelihood function is defined "likelihood_fun()" which calculates the negative log-likelihood of the GEV distribution given in the data.

The loss function "fit_distribution()" is defined to optimize the likelihood function using the "fmin()" function. It estimates the parameter of the GEV distribution ("c") that maximizes the likelihood.

Then, a synthetic dataset is generated from the GEV distribution using a known parameter "c". The parameters of the GEV distribution are estimated ("c_est") using the "fit_distribution()" function and the generated data. As a result, the true and estimated parameters are printed.

After that, a visualization (Fig. 4.4) of the observed and fitted GEV probability density functions (PDFs) and histograms of the generated data and the observed data are obtained from the estimated distribution. The results, including the observed GEV PDF and the fitted GEV PDF, are then plotted with the histograms of the generated and estimated data. At last, the plot is displayed and saved as a PDF file.

4.2.6 Inliers and Outliers in Hydrologic Data

Anomalies in data, or outliers, are values that deviate significantly from other observations in a dataset. Data anomalies can arise due to many reasons such as measurement errors, rare hydrological events, and missing data points. Anomalies can occur in univariate as well as multivariate datasets. Although they can sometimes fetch deeper insights about critical events, they can also distort analyses resulting in erroneous models and misleading conclusions. Therefore, pre-processing of the data becomes essential. Now, we present a set of statistical techniques for addressing anomalies in data.

In the context of hydrology data, the data points can be either categorized as inliers or outliers. Inliers are defined as errors in observations that lie within the expected range based on an assumed statistical model. On the other hand, outliers are data points that exhibit significant deviations from the rest of the observations. For example, outliers are those data points that correspond to extreme flood events in a river flow record, as they appreciably differ from the typical flow values.

It is extremely important to handle the outliers as they can have a considerable influence on the statistical properties of the dataset like the mean and standard deviation. Not handling them can have a detrimental effect on the performance of the data-driven models.

However, it is essential to determine whether an outlier represents an error or a true but extreme observation. Genuine outliers can provide valuable insights into the behavior of hydrological systems under extreme conditions and should be carefully considered in the analysis. It is important to remember that the classification of data points as inliers or outliers often depends on the context and the question under investigation.

4.2.7 Missing Data

Missing data is a valid problem in hydrological studies, and is capable of compromising the integrity of analyses and the robustness of model predictions. Hydrological data can have missing values due to various reasons, including equipment failure, data processing errors, or gaps in data collection efforts. This issue is especially prevalent in historical data and data collected from remote or hard-to-access areas.

Missing data hinders conducting a comprehensive analysis or accurately modeling hydrological processes. Crucial information may be lost, and the temporal continuity of datasets may be disrupted, leading to inaccurate representation of seasonal variations or long-term trends.

To manage missing data, hydrologists often employ various techniques such as interpolation, regression, or data imputation methods. However, these methods have their own limitations and can introduce additional uncertainties if not applied appropriately.

The method for handling missing data should be selected after considering the nature of the missing data, the percentage of missing values, and the underlying patterns in the data. However, preventing data loss through robust data collection, storage, and management practices remains the best approach to mitigate the impact of missing data in hydrology studies.

4.2.8 Q-Q Plots

Quantile-Quantile (Q-Q) plots are valuable graphical tools used in hydrology and other fields for assessing if a dataset follows a particular theoretical distribution. These plots compare two probability distributions by plotting their quantiles against each other. If the two distributions being compared are similar, the points in the Q-Q plot will approximately lie along the 45° line.

Though Q-Q plots are commonly used to verify the assumption of normality, they can be used for other distributions as well. For instance, a hydrologist might use a Q-Q plot to determine if rainfall or river flow data follow a normal distribution. In this case, the quantiles of the observed data would be plotted against the quantiles of a standard normal distribution.

Q-Q plots are particularly useful because they visually present the data's departure from the theoretical distribution across the full range of data, not just the mean or the median. They can highlight issues such as skewness (if the points form a curved pattern), or heavy- or light-tailed behavior (if the points fall below or above the line at the extremes of distribution).

Therefore, Q-Q plots are not only used for hypothesis testing (i.e., does the data follow a certain distribution?) but also to provide valuable insights about the nature of the distribution of the data. This information can guide the selection of appropriate statistical methods or data transformations for the subsequent analysis of the hydrological data.

Chapter 5
Graphical Hydrological Data Analysis

Hydrological relationships are often complex and nonlinear, and being able to simplify and interpret data governing such relationships is a vital component of hydrologic studies.

Identifying temporal patterns and trends is a very common form of graphical data analysis in hydrology. Seasonal and long-term variances in such datasets are easier to inspect graphically. For example, seasonal patterns or potential changes induced by climatic transitions in rainfall or river flow can be easily discerned from a time series graph. Understanding these patterns and trends visually is important before applying more robust statistical measures to them.

Another common form of graphical analysis in hydrology involves interpreting the relationship between diverse hydrological variables. Scatterplots, for instance, may be employed to uncover correlations or potential relationships between variables, such as precipitation and runoff or soil moisture and evapotranspiration. Improved understanding of these variables will facilitate building of better prediction methods

Graphical data analysis is also an effective tool for anomaly detection. Outliers or unusual events, such as floods or droughts, may not be detected through statistical summaries alone. However, by visualizing the data, these significant anomalies can be readily identified and further investigated. Detection of these anomalies is important as they can help build hydrological models and make predictions. Further, many statistical methods rest upon certain data assumptions. For example, assumptions of data normality or homoscedasticity are common in statistical modeling. Graphical methods, such as histograms, boxplots, or Q-Q plots, can be used as intuitive tools to verify these assumptions.

In Fig. 5.1 we can see an illustrative scatterplot that shows a relationship between runoff and rainfall. An upward trend is observed indicating an increase in runoff with the increase in rainfall. We can also observe some data points that clearly deviate from the trend—the outliers. They are the anomalies in the scatterplot and in the

© The Author(s), under exclusive license to Springer Nature Singapore Pte Ltd. 2024   43
A. Kumar and M. Saharia, *Python for Water and Environment*, Innovations in Sustainable
Technologies and Computing, https://doi.org/10.1007/978-981-99-9408-3_5

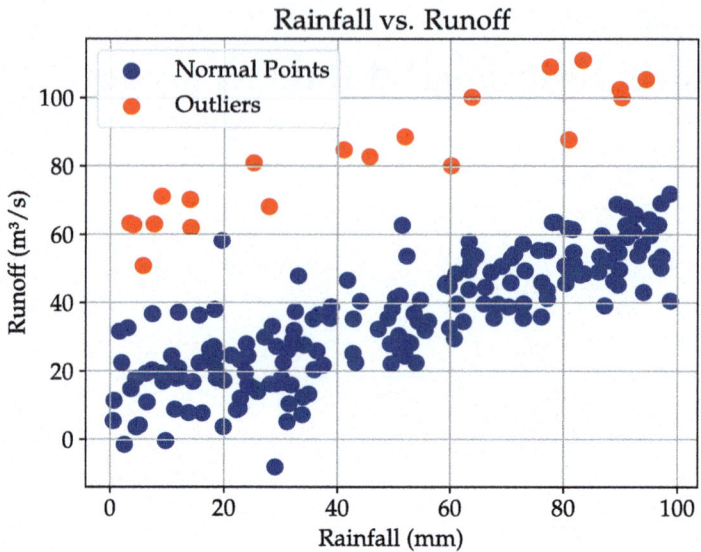

Fig. 5.1 Scatterplot showing the variation of runoff with rainfall

current context may arise due to measurement inaccuracies, localized heavy rainfall and obstructions in water flow.

Another important aspect of graphical data analysis is its ability to foster data exploration and hypothesis generation. Researchers can unearth new insights and generate fresh hypotheses from the visualized data. For example, Anscombe's quartet (Fig. 5.2) comprises four distinct datasets that have identical statistical properties (mean, variance, correlation, and linear regression coefficients), yet appear very different when graphed. It underscores the importance of graphical data analysis in addition to numerical analysis. This exploratory process often spurs further investigation and analysis, enhancing the overall depth of the study.

The capacity of graphical analysis to communicate complex data and results is unparalleled. Data visualizations can convey intricate data structures and research findings to diverse audiences, including policymakers, stakeholders, and the general public, in an understandable and meaningful way.

Graphical data analysis in hydrology complements numerical analysis, enabling a holistic understanding of the data. It greatly facilitates effective decision-making in water resource management. It must be noted that while graphical analysis is an effective tool, it must be used alongside other statistical techniques for a thorough and comprehensive analysis.

5.1   For a Single Dataset

One-dimensional (1D) datasets are frequently encountered in the form of time series data that record variables such as precipitation, temperature, streamflow, and water level over a period of time at a particular location. Other cases may involve extracting

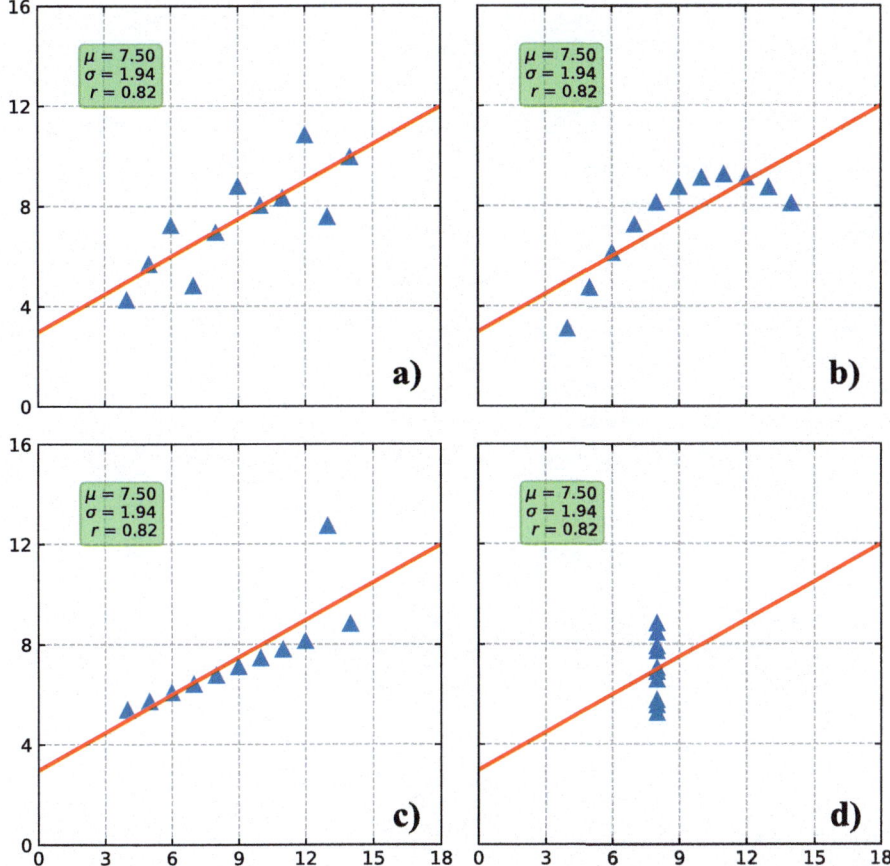

Fig. 5.2 Datasets **a–d**, with linear regression fit and scatterplots. Anscombe's quartet comprises four datasets that have nearly identical and simple statistical properties (mean, variance, correlation, and linear regression line), yet appear very different when graphed. Each dataset consists of 11 (x, y) points. They were constructed in 1973 by statistician Francis Anscombe to demonstrate the importance of graphing before data analysis and the effect of outliers on statistical properties

single columns from multivariate data. Below are some graphical tools that can be used for 1D datasets.

5.1.1 Histograms

A histogram is a graphical representation of data distribution. It uses bars of different heights to illustrate the frequency of data points in different ranges or bins. The x-axis represents the bins, and the y-axis represents the frequency. It gives a clear and comprehensive visualization and understanding of the statistical properties of numerical data.

```python
"""
Histogram for single dataset
"""


# Import libraries
import pandas as pd
import seaborn as sns
import matplotlib.pyplot as plt



"""
Load dataset
"""
data = pd.read_csv(
    filepath_or_buffer="../data/Godavari.csv",
    sep=",",
    header=0).dropna()
print("\nChecking data:")
try:
    data["time"] = pd.to_datetime(
        data['time'], infer_datetime_format=True)
    print("    Date format is okay!\n")
except ValueError:
    print("    Encountered error!\n")
    pass

Level_data = data[["Level"]]
del data
print("Read data file")



"""
Plotting the histogram
"""
sns.histplot(data=Level_data,
             bins=50)
plt.xlabel(Level_data.columns[0] + " (m)")
plt.grid(ls='--')
plt.tight_layout()
plt.show()
plt.savefig("single_hist.pdf", dpi=300)
print("\n\nPlotted!")
```

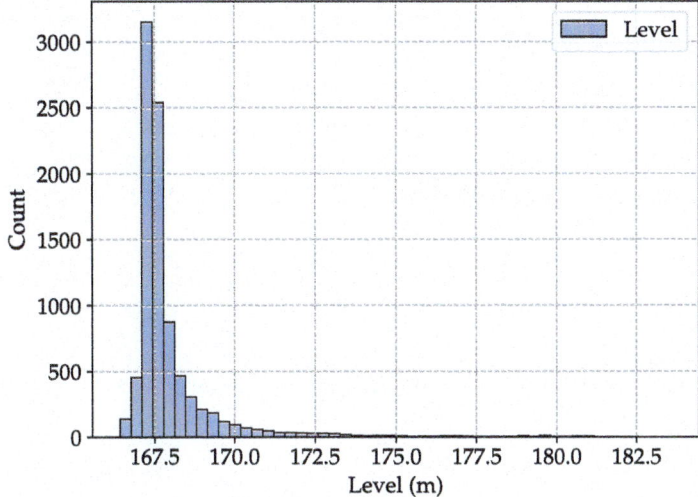

Fig. 5.3 Histogram of daily Mean Water Level (m) for the Godavari River from 1981 to 2005

The given Python program is used to generate a histogram from a dataset, which, in this case, is named "Godavari.csv". The key components of the program can be broken down as follows:

At first, we import the necessary libraries for data manipulation and visualization: pandas for data manipulation combined with Seaborn, and matplotlib for data visualization.

Next, we load the CSV file using pandas' "read_csv()" function. The program then uses the "dropna()" function to drop rows with missing values, convert the "time" column to datetime format using "pd.to_datetime()", and finally keep only the "Level" column of the data for further analysis.

Then, Seaborn's "histplot()" function is used to create a histogram of the "Level" data. The histogram has 50 bins ("bins=50"). The grid lines and layout adjustments are done using matplotlib functions ("grid", "tight_layout").

Lastly, a histogram is displayed on the screen with "plt.show()" and is also saved as a PDF file using "plt.savefig()".

This Python program reads a time series dataset, processes the data to keep only the required column (in this case "Level"), and creates a histogram of the data to visualize its distribution. The result is shown in Fig. 5.3.

5.1.2 Boxplots

A boxplot, also known as a box-and-whisker plot, is a graphical representation used to describe the distribution of data. It displays a summary of the minimum, first quartile (Q1), median, third quartile (Q3), and maximum values of a dataset. The "box" represents the interquartile range (Q3–Q1), containing the middle 50% of the data, while the "whiskers" extend to the smallest and largest observations. Outliers can be indicated as individual points beyond the whiskers.

```
1   """
2   Boxplot for single dataset
3   """
4
5   # Import libraries
6   import pandas as pd
7   import seaborn as sns
8   import matplotlib.pyplot as plt
9
10
11  """
12  Load dataset
13  """
14  data = pd.read_csv(
15      filepath_or_buffer="../data/Godavari.csv",
16      sep=",",
17      header=0).dropna()
18  print("\nChecking data:")
19  try:
20      data["time"] = pd.to_datetime(
21          data['time'], infer_datetime_format=True)
22      print("    Date format is okay!\n")
23  except ValueError:
24      print("    Encountered error!\n")
25      pass
26
27  Level_data = data[["time", "Level"]]
28  del data
29  print("Read data file")
30
31  """
32  Resample:
33  Downsample the time series
34  """
35  Level_data = Level_data.resample('1M', on="time").mean()
36
37
38  """
39  Plotting the boxplot
40  """
41  sns.boxplot(data=Level_data,
42              notch=True, showcaps=False,
43              flierprops={"marker": "x"},
44              boxprops={"facecolor": (.4, .6, .8, .5)},
45              medianprops={"color": "coral"},
46              )
47  plt.ylabel("(m)")
48  plt.grid(ls='--')
49  plt.tight_layout()
50  plt.show()
51  plt.savefig("single_box.pdf", dpi=300)
52  print("\n\nPlotted!")
```

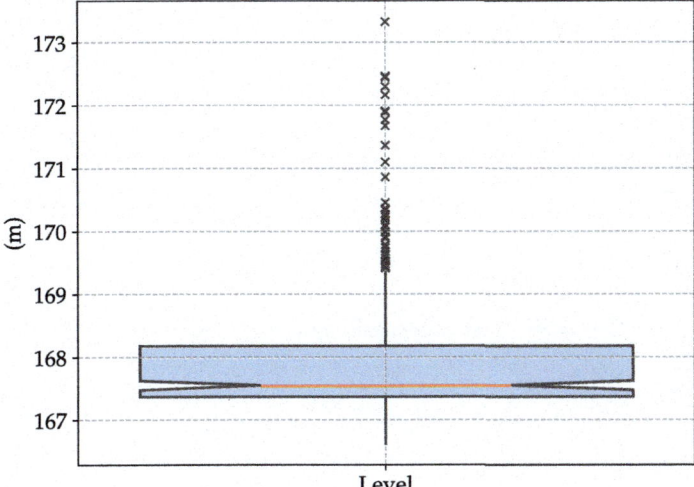

Fig. 5.4 Boxplot of monthly averaged mean water level for the Godavari River from 1981 to 2005

This Python program visualizes a dataset of water level through a boxplot using Seaborn and matplotlib libraries. The initial part of the script involves importing the necessary libraries and loading the dataset from a CSV file. The loaded data is then cleaned by removing any missing values ("dropna()"), and ensuring the "time" column is in the correct datetime format using pandas' "to_datetime()" function.

After some cleanup, the data of interest is selected—specifically, columns for "time" and "Level". The rest of the data is deleted to save memory. The data is then downsampled using the "resample()" function to calculate monthly averages, reducing the granularity of the dataset to a manageable size, and potentially revealing longer-term trends.

In the next part of the script, a boxplot is created with Seaborn's "boxplot()" function. Here, various visual adjustments are made, including notching the boxplot, hiding the caps, marking outliers with "x" symbols, adjusting colors, and adding a grid. The boxplot shows the distribution of water levels across months, providing insights into their central tendency and spread. Lastly, the plot is saved as a high-resolution PDF file.

This script demonstrates a workflow for loading, pre-processing, and visualizing time series data, with a focus on creating a detailed boxplot (Fig. 5.4) to summarize the distribution of the data.

5.1.3 Quantile Plots

Quantile plots are graphical tools used for assessing if a dataset follows a certain theoretical distribution. They plot the quantiles of the data against the quantiles of a chosen theoretical distribution. If the data aligns closely with the theoretical line, it means that the data follows that distribution.

```
1    """
2    Quantile plot for single dataset
3    """
4
5    # Import libraries
6    import pandas as pd
7    from scipy import stats
8    import pylab
9    import matplotlib.pyplot as plt
10
11
12   """
13   Load dataset
14   """
15   data = pd.read_csv(
16       filepath_or_buffer="../data/Godavari.csv",
17       sep=",",
18       header=0).dropna()
19   print("\nChecking data:")
20   try:
21       data["time"] = pd.to_datetime(
22           data['time'], infer_datetime_format=True)
23       print("   Date format is okay!\n")
24   except ValueError:
25       print("   Encountered error!\n")
26       pass
27
28   Level_data = data[["time", "Level"]]
29   del data
30   print("Read data file")
31
32
33   """
34   Resample:
35   Downsample the time series
36   """
37   Level_data = Level_data.resample(
38       rule='1M', on="time").mean()
39
40
41   """
42   Plotting the q-q plot
43   """
44   stats.probplot(
45       x=Level_data['Level'],
46       dist="norm",
47       plot=pylab
48   )
49   plt.xlabel("Sample quantiles")
50   plt.ylabel("Ranked Level data (m)")
51   plt.title("")
52   plt.grid(ls='--')
53   plt.axis('square')
54   plt.xlim(-4.5, 4.5)
55
56   plt.tight_layout()
57   plt.show()
58   plt.savefig("single_qq.pdf", dpi=300)
59   print("\n\nPlotted!")
```

The above Python program reads a time series dataset, resamples it to a lower frequency (monthly), and then generates a Quantile-Quantile (Q-Q) plot. Below is a step-by-step explanation of the program.

At first, the necessary Python libraries are imported. These include pandas for data manipulation, scipy's stats module for statistical functions, matplotlib's pylab module for plotting, and matplotlib.pyplot for additional plotting features.

Next, the pandas function read_csv() is used to load a CSV file named "Godavari.csv" located in the "data" folder. Rows containing missing values are discarded using the "dropna()" function. The dataset has a "time" column representing timestamps and a "Level" column representing the water level data.

The program then converts the "time" column into a standard pandas Datetime object. This is necessary for time-based functionalities available in pandas.

After loading and checking the data, the "time" and "Level" columns are extracted and saved as "Level_data". The original dataframe "data" is then deleted to save memory.

Then, the "Level_data" time series is downsampled to a monthly frequency using "resample()" function in pandas. The "mean()" function is then applied to calculate average "Level" for each month.

The "probplot()" function from scipy's stats module is then used to generate the Q-Q plot. This plot compares the quantiles of the "Level" data to the quantiles of a normal distribution.

Labels for the x-axis and y-axis are added, the grid is displayed, and the aspect ratio of the plot is set to "square" with limits from −4.5 to 4.5 on both axes. The plot layout is then adjusted using "tight_layout()" and displayed with "show()". The plot is also saved as "single_qq.pdf" at a resolution of 300 dpi.

This script, thus, provides a complete workflow for loading, preprocessing, and visualizing a time series dataset, specifically focusing on testing the normality of the data through a Q-Q plot as shown in Fig. 5.5.

Fig. 5.5 Quantile-quantile plot of monthly averaged mean water level for the Godavari River data. The x-axis represents the quantiles where the data would fall assuming a Gaussian distribution on the data

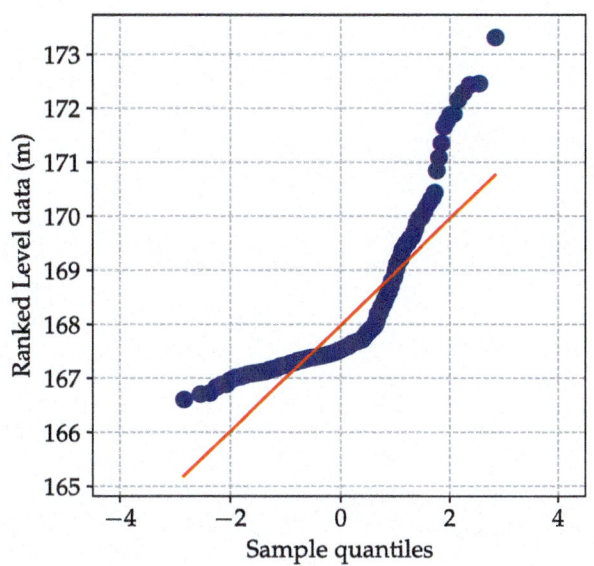

5.2   For Multivariate Data

Multivariate data is often encountered in hydrological studies. Its analysis is critical for understanding complex hydrological processes and interactions. Below we discuss a few graphical tools that can be applied in analyzing these types of data.

5.2.1   Scatter Matrix Plot

A scatter matrix plot, or pair plot, is a graphical tool that presents scatterplot for every pair of features and histograms along the diagonal in a dataset. It allows for a quick visual examination of potential relationships or patterns between variables. This tool is useful for exploring data and identifying correlations, clusters, outliers, or trends across multiple dimensions in a dataset. It provides a way to visualize high-dimensional data on a two-dimensional graph.

```
 1   """
 2   Program to demonstrate scatter
 3    matrix using seaborn
 4   """
 5
 6   # Import libraries
 7   import pandas as pd
 8   import matplotlib.pyplot as plt
 9   import seaborn as sns
10
11
12   """
13   Loading the dataset
14   """
15   data1 = pd.read_csv( # Loading first dataset
16       filepath_or_buffer="../data/Godavari.csv",
17       sep=",",
18       header=0
19   ).dropna()
20
21   df1 = data1[[
22       "Level", "Streamflow",
23       "Pressure", "Rel_humidity"]]   # Retrieving columns
24   df1.loc[:, "River"] = "Godavari"   # Creating new column
25   df1 = df1.iloc[:2000, :]
26   data2 = pd.read_csv( # Loading second dataset
27       filepath_or_buffer="../data/Cauvery.csv",
28       sep=",",
29       header=0
30   ).dropna()
31   df2 = data2[[
32       "Level", "Streamflow",
```

```
1        "Pressure", "Rel_humidity"]]   # Retrieving columns
2   df2.loc[:, "River"] = "Cauvery"   # Creating new column
3   df2 = df2.iloc[:2000, :]
4
5   # Combining dataset
6   df = pd.concat([df1, df2])
7
8
9   """
10  Visualization
11  """
12  sns.color_palette("Set2")
13  sns.pairplot(df, hue="River", height=2.0, markers='x')
14  plt.show()
15  plt.savefig("multi_scatter.pdf", dpi=300)
```

The above Python program creates a scatter matrix visualization using two datasets, one for the Godavari River and the other for the Cauvery River. This visualization allows the examination of the relationships among different variables from the datasets.

To begin with, the necessary libraries are imported—pandas for data manipulation, matplotlib.pyplot for basic plotting, and Seaborn for advanced data visualization.

Next, the program loads the first dataset pertaining to the Godavari River. It reads the dataset from a CSV file named "Godavari.csv" and removes any rows with missing values using the "dropna()" function. It then selects a subset of columns ("Level", "Streamflow", "Pressure", "Rel_humidity") for analysis and adds a new column "River", filling it with the value "Godavari". This process is repeated for the second dataset pertaining to the Cauvery River, with the new column "River" being filled with the value "Cauvery". In both cases, only the first 2000 rows of the dataset are considered for further analysis.

Once both datasets are prepared, they are combined using the "concat()" function from pandas. This results in a single dataframe containing data from both rivers.

Finally, the program visualizes the combined dataset using a scatter matrix. The scatter matrix is a pairplot generated by Seaborn's "pairplot()" function, with different colors representing different rivers as specified by the "hue" argument. The "height" argument sets the size of the plot and "markers" specifies the marker style. After generating the plot, it is displayed using "plt.show()" as shown in Fig. 5.6. The plot is also saved as a PDF file with a resolution of 300 dpi. The color palette for the plot is set to "Set2" using Seaborn's "color_palette()" function.

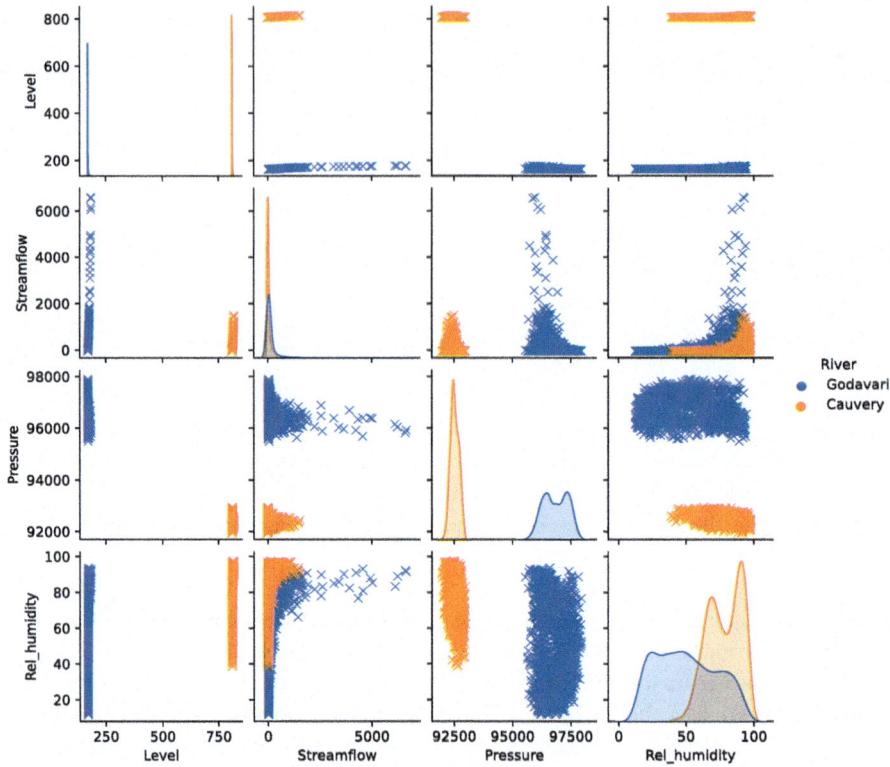

Fig. 5.6 Scatter matrix plot for various properties of two rivers—Godavari and Cauvery

5.2.2 *Parallel Coordinate Plot*

A parallel coordinate plot is a visualization tool for multi-dimensional data. Each variable in the dataset is represented by a vertical axis, and the data points are displayed as lines crossing these axes. The position of the line on each axis represents the value of the corresponding variable. It is particularly useful for exploring relationships and patterns across several variables simultaneously.

```
1   """
2   Program to demonstrate parallel
3   plots using pandas
4   """
5
6   # Import libraries
7   import pandas as pd
8   import matplotlib.pyplot as plt
9   from pandas.plotting import parallel_coordinates
10
11
12  """
13  Loading the dataset
14  """
15  data1 = pd.read_csv( # Loading first dataset
16      filepath_or_buffer="../data/Godavari.csv",
17      sep=",",
18      header=0
19  ).dropna()
20  df1 = data1[[
21      "Level", "Streamflow",
22      "Pressure", "Rel_humidity"]]   # Retrieving columns
23  df1.loc[:, "River"] = "Godavari"   # Creating new column
24  df1 = df1.iloc[:2000, :]
25
26  data2 = pd.read_csv( # Loading second dataset
27      filepath_or_buffer="../data/Cauvery.csv",
28      sep=",",
29      header=0
30  ).dropna()
31  df2 = data2[[
32      "Level", "Streamflow",
33      "Pressure", "Rel_humidity"]]   # Retrieving columns
34  df2.loc[:, "River"] = "Cauvery"   # Creating new column
35  df2 = df2.iloc[:2000, :]
36
37  df = pd.concat([df1, df2])
38
39
40  """
41  Visualization
42  """
43  parallel_coordinates(
44      frame=df, class_column='River',
45      colormap=plt.get_cmap("Set2")
46  )
47  plt.grid(ls='--')
48  plt.tight_layout()
49  plt.show()
50  plt.savefig("multi_parallel.pdf", dpi=300)
```

The above program starts by importing the necessary libraries. Here, pandas is used for data manipulation and analysis, matplotlib.pyplot for visualizations in

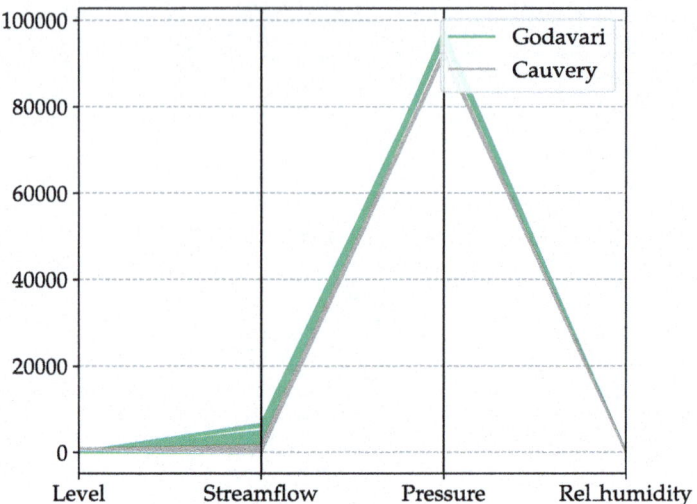

Fig. 5.7 Parallel coordinate plot for various properties of the two rivers—Godavari and Cauvery

Python, and the parallel_coordinates() function from pandas.plotting for creating parallel plots.

The program then loads the first dataset from a CSV file named "Godavari.csv", using the pandas function pd.read_csv(). The loaded data is cleaned by dropping any row that contains missing values with the .dropna() function. Only the "Level", "Streamflow", "Pressure", and "Relative Humidity" columns are selected from the dataset, and a new column "River" is added with a constant value "Godavari". The process is repeated for the second dataset, "Cauvery.csv", with the corresponding river name added to its "River" column. For both datasets, only the first 2000 rows are used for visualization.

Afterward, both datasets are concatenated into a single pandas dataframe using the pd.concat() function. This newly combined dataframe now holds data from both the Godavari and Cauvery Rivers.

Finally, the visualization is created using the parallel_coordinates() function, which plots each feature on a separate column and then draws lines connecting the features for each data sample. In this case, the "River" column is used as the class column, meaning that separate lines will be drawn and color-coded for each river. The colormap is set to "Set2" which is a qualitative colormap suitable for distinguishing different categories. The plt.grid() function is used to add gridlines to the plot, and plt.tight_layout() adjusts the subplot params so that the subplot fits into the figure area.

The plot is displayed using plt.show() as shown in Fig. 5.7. It is then saved to a PDF file named "multi_parallel.pdf" with a resolution of 300 dpi using plt.savefig(). This program is a handy way to visualize and compare the multivariate data from the two rivers.

5.3 Publication-Ready Graphics

Creating publication-ready graphics involves much more than simply representing data visually. It requires careful design choices and attention to detail to ensure clarity, accuracy, and relevance. Below are some characteristics of good publication-ready graphics:

1. Clear and Concise: A good graphic should convey the main message quickly and clearly. Overly complex or cluttered graphics can confuse readers and obscure key points. The inclusion of data and elements should be deliberate, with the aim to enhance the understanding of the graphic's central theme.
2. Accurate and Honest Representation: Data should be represented accurately without any manipulations which could mislead readers. This includes proper scaling of axes, appropriate use of data points, and avoiding distortions or exaggerations of data patterns.
3. Labels and Titles: Every graphic should have a clear and descriptive title that outlines what it represents. Labels for axes, data series, or other graphic elements are also crucial for comprehension. Legends or keys should be provided where necessary.
4. Appropriate Use of Colors and Symbols: Colors and symbols used should enhance the clarity of the graphic, not detract from it. Colors should be distinguishable but not overwhelming, and they should consider color-blind readers. Symbols should be clearly differentiated and must be consistent throughout the graphic.
5. Data Source and Units: Publication-ready graphics should always indicate the source of data and the units of measurement. This information is typically placed in the caption or on the axis labels.
6. Consistency: If multiple graphics are used in the same publication, they should maintain a consistent style, including typography, color palette, and usage of symbol. This provides a cohesive look and feel to the publication and makes it easier for readers to compare and relate the graphics.
7. Simplicity: Simplicity is the key to good graphics. A simple, clean design helps the reader focus on the data and message it conveys. One should avoid unnecessary embellishments that do not contribute to understanding the data.
8. High Resolution: Finally, the graphic must be high resolution to ensure that it is clearly visible, both on-screen and in print. It should not appear pixelated or blurry, and all text should be legible.

When the above characteristics are met, the graphics used in the publications can effectively communicate the intended message, enhancing the reader's understanding, and adding visual appeal to the publication. The goal should always be to make complex data understandable and accessible to a wide audience.

5.4 Misleading Graphics

Misleading graphics in the field of hydrology, are visual representations of data or results that can lead to false interpretations or misconceptions. While at times these inaccuracies are unintentional, stemming from poor design choices, at other times, they might be deliberately employed to skew perceptions and advance a particular narrative. Regardless of the intent, misleading graphics can undermine the integrity of research and lead to erroneous conclusions.

In hydrology, graphics often portray complex data related to water cycles, climate change, water quality, or other hydrological phenomena. These might include line graphs, bar charts, pie charts, maps, or more intricate visualizations. There are several ways to create misleading graphics in hydrology.

One common form is inappropriate scale manipulation. This involves adjusting the y-axis (or x-axis) to make changes or differences appear larger or smaller than they are. For instance, a graph showing a slight increase in river pollution levels over time might use a truncated y-axis to make the increase seem more drastic.

Another form is cherry-picking data, which involves selecting and displaying only data that supports a certain conclusion while ignoring the rest. In hydrology, this might mean selectively presenting data from years with high rainfall to suggest an upward trend in precipitation, while neglecting years with low rainfall.

Misrepresentation of spatial data on maps is also frequently issued. This could involve using inappropriate color scales, distorting geographical areas, or neglecting to provide a key for interpretation.

Sometimes, complex hydrological phenomena can be oversimplified, leading to the loss of crucial information. For example, representing a multidimensional water quality index with a single color or number might oversimplify the situation and overlook crucial details.

Misleading graphics can significantly distort scientific communication in hydrology, leading to misinterpretations and misinformation. As such, it is vital for researchers, reviewers, and readers to keep a critical eye when creating or interpreting graphics, to ensure the integrity and accuracy of scientific research in hydrology.

Part II
Statistical Modeling in Hydrology

Chapter 6
Curve Fitting and Regression Analysis

Curve fitting and regression analysis are powerful statistical tools used widely in hydrological data modeling. They can be used to model underlying relationships between data, allowing you to interpret and predict hydrological behavior under varying conditions.

Curve fitting involves fitting a function on a set of data points that best represents the underlying trend in the dataset. Curve fitting supports essential tasks such as deriving intensity-duration-frequency (IDF) curves for rainfall, which describe the occurrence frequency, duration, and intensity of a rain event. These curves are critical for managing flood risk and designing effective dams, reservoirs, or drainage systems. Likewise, they also help in modeling hydrographs, a way to depict the water flow rate over time, and anticipating and managing river discharge patterns.

Regression analysis, a form of curve fitting, is often used to model the relationship between two or more variables. This technique determines the equation that best describes the dependent variable in terms of the independent variables. Rainfall-runoff modeling can use regression analysis to establish the relationship between streamflow (dependent variable) and rainfall (independent variable). This relationship is vital for water resource management, water availability, and forecasting of floods. It also finds applications in estimating the groundwater recharge rates from the rate of recharge and depletion over time and in pumping tests for drawndown curve modeling and estimating important aquifer parameters such as hydraulic conductivity.

Analysis of water quality parameters is another application of regression analysis in environmental engineering. It can help establish relationships between concentrations of pollutants and factors like flow rate, temperature, and land-use characteristics.

© The Author(s), under exclusive license to Springer Nature Singapore Pte Ltd. 2024
A. Kumar and M. Saharia, *Python for Water and Environment*, Innovations in Sustainable Technologies and Computing, https://doi.org/10.1007/978-981-99-9408-3_6

These relationships can influence processes of water treatment and water pollution controlling strategies.

Curve fitting and regression analysis play a crucial role in the study of the impacts of climate change on hydrological processes. It allows hydrologists to relate hydrological variables such as evapotranspiration, precipitation, and runoff to climate variables like temperature and carbon dioxide concentrations. These models also become essential in predicting future water availability and adaptation strategies under changing climate scenarios.

Although these techniques are immensely beneficial, it is essential to note that they are based on the assumption that the relationships will also hold in the future. Therefore, one should appropriately address the associated uncertainties in respective studies.

Curve fitting and regression analysis are invaluable in hydrology, offering insights into complex hydrological processes, influencing water resource management, and helping predict future scenarios. By uncovering patterns and relationships in historical data, these tools contribute significantly to our understanding and stewardship of water systems. In the following sections, we describe the simple, multiple, and nonlinear regression techniques.

6.1   Simple Linear Regression of Flow

Simple linear regression is a common statistical method for estimating the relationship between one dependent and one independent variable. The dependent variable is sometimes called the target and the independent variable is called the covariate, a feature, or a predictor. A linear regression model is given by the following equation:

$$y = \beta_0 + \beta_1 x_1 + \beta_2 x_2 \cdots + \beta_p x_p + \epsilon \qquad (6.1)$$

where β_j denotes the coefficients of the predictors, x_j, and ϵ is the associated random error independent of the predictors. The aim of linear regression is to estimate the coefficients β_j using the method of least squares.

In the associated program, we build a regression model between streamflow and water level for the Godavari basin data. The regression equation for this case can be given as

$$\text{Streamflow} = \beta_0 + \beta_1 * \text{Water Level} \qquad (6.2)$$

```python
"""
Program to do simple linear
regression on the Godavari
streamflow data
"""

# Import libraries
import pandas as pd
import numpy as np
from sklearn.linear_model import LinearRegression
from sklearn.model_selection import train_test_split
import matplotlib.pyplot as plt
import seaborn as sns


savePlots = 1


"""
Load dataset
"""
data = pd.read_csv(
    filepath_or_buffer="../data/Godavari.csv",
    sep=",",
    header=0).dropna()
print("\nChecking data:")
try:
    data["time"] = pd.to_datetime(
        data['time'], infer_datetime_format=True)
    print("   Date format is okay!\n")
except ValueError:
    print("   Encountered error!\n")
    pass
df = data[["time", "Pressure",
           "Rel_humidity", "Level", "Streamflow"]]
del data
print("Read data file")

df.head()


"""
Data preprocessing
"""
```

```
45  """
46  Linear Regression model fitting
47  """
48  linearRegression = LinearRegression()
49  linearRegression.fit(X_train, y_train)
50  b0, b1 = linearRegression.intercept_, \
51      linearRegression.coef_[0]
52  print("\n\nEstimated parameters: "
53      "\n    (b0, b1) = ({:.2f}, {:.2f})".
54      format(b0, b1))
55  print("R-squared value: \n    R^2 = {:.2f}".
56      format(linearRegression.score(X_train, y_train)))
57
58
59  """
60  Visualization of results
61  """
62  # Plot the datapoints
63  sns.scatterplot(x=np.squeeze(X_train), y=y_train)
64  plt.grid(ls="--")
65  plt.xlabel("Level(m)")
66  plt.ylabel("Streamflow(cumecs)")
67
68  # Plot regression line
69  x1, x2 = np.min(X), np.max(X)
70  y1, y2 = (b1*x1 + b0), (b1*x2 + b0)
71  plt.plot([x1, x2], [y1, y2], 'r-', label="Linear Model")
72  plt.legend()
73  plt.grid(ls="--")
74
75  # Tidying the plot
76  plt.tight_layout()
77  plt.show()
78  plt.savefig("linearReg_.pdf", dpi=300)
79
80  print("Done!!")
```

```
●  ●  ●            Terminal

>>>          time  Pressure  Rel_humidity   Level  Streamflow
>>> 0 1981-01-01  97416.18     60.102580  167.60        10.5
>>> 1 1981-01-02  97340.45     58.449010  167.59        10.3
>>> 2 1981-01-03  97306.44     58.281210  167.59        10.2
>>> 3 1981-01-04  97270.30     57.049640  167.59        10.4
>>> 4 1981-01-05  97371.98     57.171913  167.58        10.0
>>>
>>> Estimated parameters:
>>>    (b0, b1) = (-41624.79, 248.61)
>>> R-squared value:
>>>    R^2 = 0.91
```

The given program performs simple linear regression on the Godavari stream-flow data. It uses several libraries, namely pandas, NumPy, sklearn, matplotlib, and Seaborn for data manipulation, mathematical operations, modeling, and data visualization.

The first stage of the program involves loading the dataset. The data is read from a CSV file named "Godavari.csv" using pandas" "read_csv()" function and any rows with missing data are removed. An attempt is then made to convert the "time" column of the dataset into a datetime format using pandas" "to_datetime()" function. The data of interest, i.e., the "time", "Pressure", "Rel_humidity", "Level", and "Streamflow" columns, are extracted into a new dataframe "df".

The program then enters the data preprocessing stage. Here, the "Level" column (the independent variable) is reshaped into a 2D array to be used as the feature (X) for the linear regression model. The "Streamflow" column (the dependent variable) is the target (y). This data is split into training and testing sets using the "train_test_split()" function from sklearn's "model_selection" module, with 33% of the data held out for testing. It may be noted that the "train_test_split()" function splits the data randomly as per the given ratio "test_size". However, for the retrievability of the results a "random_state" is defined. The results may vary for a different value of "random_state".

Following data preprocessing, the linear regression model is instantiated and fitted using the training data. The intercept and coefficient of the linear model are printed, along with the R-squared value, a measure of how well the linear model fits the data.

Finally, the results are visualized using matplotlib and Seaborn libraries. A scatterplot of "Level" versus "Streamflow" is created, on top of which the best-fit line generated by the linear regression model is superimposed. The resulting figure is displayed (Fig. 6.1) and then saved as a PDF file named "linearReg_.pdf" for future reference.

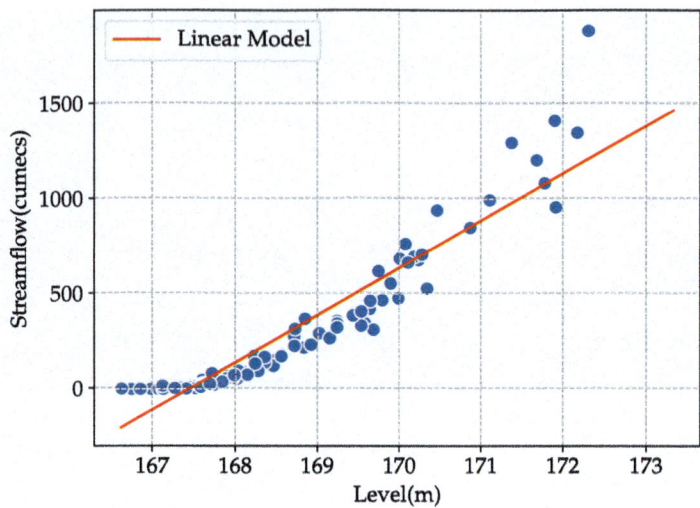

Fig. 6.1 Linear regression plot of the Streamflow versus Water Level of the Godavari River data. The data is not well-fitted with the current linear model

6.2   Multiple Linear Regression of Flow

We now describe a multiple linear regression case using the following program. In the following program, we create a regression model to predict the streamflow as a function of level and relative humidity. The equation for multiple linear regression can be given as

$$\text{Streamflow} = \beta_0 + \beta_1 * \text{Rel\_humidity} + \beta_2 * \text{Level} \qquad (6.3)$$

```
1   """
2   Program to do multiple linear
3   regression on the Godavari
4   streamflow data
5   """
6
7   # Import libraries
8   import pandas as pd
9   import numpy as np
10  from sklearn.linear_model import LinearRegression
11  from sklearn.model_selection import train_test_split
12  import matplotlib.pyplot as plt
```

```
1  """
2  Load dataset
3  """
4  data = pd.read_csv(
5      filepath_or_buffer="../data/Godavari.csv",
6      sep=",",
7      header=0).dropna()
8  print("\nChecking data:")
9  try:
10     data["time"] = pd.to_datetime(
11         data['time'], infer_datetime_format=True)
12     print("   Date format is okay!\n")
13 except ValueError:
14     print("   Encountered error!\n")
15     pass
16 df = data[["time", "Pressure",
17           "Rel_humidity", "Level", "Streamflow"]]
18 del data
19 print("Read data file")
20
21 """
22 Resample:
23 Downsample the time series
24 """
25 df = df.resample('1M', on="time").mean()
26
27 """
28 Data preprocessing
29 """
30 X = df[["Pressure", "Level"]].values.reshape(-1, 2)
31 y = df["Streamflow"].values
32 X_train, X_test, y_train, y_test = train_test_split(
33     X, y, test_size=0.33, random_state=11)
34
35 """
36 Linear Regression model fitting
37 """
38 multipleRegression = LinearRegression()
39 multipleRegression.fit(X_train, y_train)
40 b0, b1, b2 = multipleRegression.intercept_, \
41     multipleRegression.coef_[0], multipleRegression.coef_[1]
42 print("\n\nEstimated parameters: "
43     "\n   (b0, b1, b2) = ({:.2f}, {:.2f}, {:.2f})".
44     format(b0, b1, b2))
45 print("R-squared value: \n    R^2 = {:.2f}".
46     format(multipleRegression.score(X_train, y_train)))
47
48 """
49 Visualization of results
50 """
```

```
39  # Extract min max
40  xmin, xmax = np.min(X_train[:, 0]), np.max(X_train[:, 0])
41  ymin, ymax = np.min(X_train[:, 1]), np.max(X_train[:, 1])
42  zmin, zmax = np.min(y_train), np.max(y_train)
43
44  # Surface equation
45  XX = np.linspace(xmin, xmax, 20)
46  YY = np.linspace(ymin, ymax, 20)
47  xx, yy = np.meshgrid(XX, YY)
48  zz = b0 + b1 * xx + b2 * yy
49
50  # Surface plot and Scatter 3d plot of points
51  fig = plt.figure(figsize=(7, 6))
52  ax = plt.axes(projection='3d')
53  ax.plot_surface(xx, yy, zz,
54                  rstride=1, cstride=1, alpha=0.5)
55  ax.scatter(X_train[:, 0], X_train[:, 1], y_train,
56             marker='o', edgecolors='black', c='red', s=30)
57
58  # Tick numbers and locations
59  xtick_loc = np.linspace(xmin, xmax, 5)
60  ytick_loc = np.linspace(ymin, ymax, 5)
61  ztick_loc = np.linspace(zmin, zmax, 5)
62  ax.set_xticks(xtick_loc)
63  ax.set_yticks(ytick_loc)
64  ax.set_zticks(ztick_loc)
65
66  # Labels
67  ldist = 9
68  ax.set_xlabel('Relative Humidity (X)', labelpad=ldist)
69  ax.set_ylabel('Level (Y)', labelpad=ldist + 4)
70  ax.set_zlabel('Streamflow (Z)', labelpad=ldist)
71
72  # Tidying up
73  ax.view_init(elev=25, azim=-50)
74  ax.axis('auto')
75  plt.tight_layout()
76  plt.show()
77
78  # Saving
79  plt.savefig("multipleReg_.pdf", dpi=300)
80  print("Done!!")
```

```
●  ●  ●        Terminal

   Estimated parameters:
      (b0, b1, b2) = (-47913.28, -3.21, 287.03)
   R-squared value:
      R^2 = 0.93
```

The script loads a CSV data file named "Godavari.csv" into a pandas dataframe and drops any missing values. It attempts to convert the "time" column to the date-

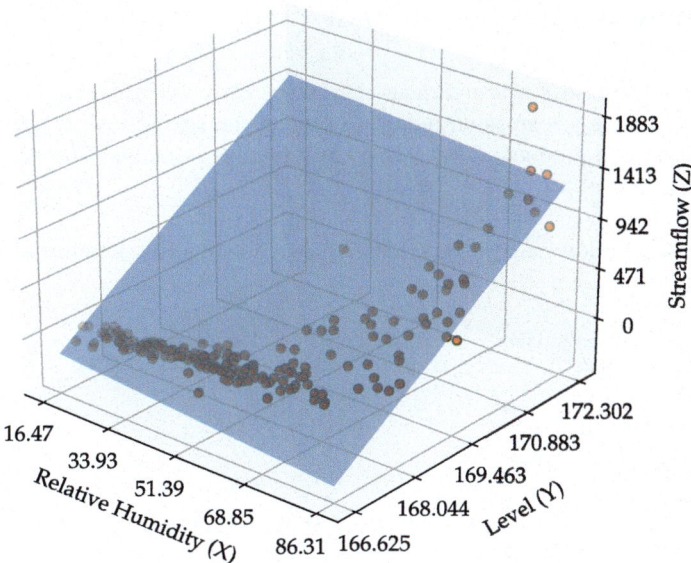

Fig. 6.2 Multiple linear regression plot of the Streamflow versus Level and Relative_humidity data of the Godavari River

time format and then selects the "time", "Pressure", "Rel_humidity", "Level", and "Streamflow" columns for analysis. The dataframe is then downsampled to a monthly timescale using the pandas resample() function, which takes the mean of all data points within each month on the "time" column.

After data loading and preprocessing, the script extracts the "Rel_humidity" and "Level" columns as input features (X), and the "Streamflow" column as the target variable (y). It then splits these into training and test sets, with 67% of data allocated for training and the rest for testing.

Following this, the script initializes a LinearRegression object from sklearn, fits the model using the training data, and prints the estimated parameters (intercepts and coefficients) of the linear model. It also calculates the R-squared score using the training data, which measures the goodness of fit of the model.

Finally, the script visualizes the results in a 3D plot (Fig. 6.2), where the "Rel_humidity" and "Level" data form the x and y-axes, respectively, and the predicted "Streamflow" forms the z-axis. The actual training data points are overlaid on this surface as a scatterplot.

6.3   Nonlinear Regression of Flow

This section demonstrates a nonlinear regression code example. In the following program, we create a regression model to predict the streamflow as a function of water level and the product of water level and relative humidity. The equation can be given as follows:

$$\text{Streamflow} = \beta_0 + \beta_1 * \text{Level} + \beta_2 * \text{Level} * \text{Rel\_humidity} \qquad (6.4)$$

```python
"""
Program to do nonlinear regression on the Godavari
streamflow data
"""


# Import libraries
import pandas as pd
import numpy as np
from sklearn.linear_model import LinearRegression
from sklearn.model_selection import train_test_split
import matplotlib.pyplot as plt


"""
Load dataset
"""
data = pd.read_csv(
    filepath_or_buffer="../data/Godavari.csv",
    sep=",",
    header=0).dropna()
print("\nChecking data:")
try:
    data["time"] = pd.to_datetime(
        data['time'], infer_datetime_format=True)
    print("   Date format is okay!\n")
except ValueError:
    print("   Encountered error!\n")
    pass
df = data[["time", "Pressure",
           "Rel_humidity", "Level", "Streamflow"]]
del data
print("Read data file")


"""
Resample:
Downsample the time series
"""
df = df.resample('1M', on="time").mean()


"""
Data preprocessing
"""
```

```
43   X1 = df[["Level"]].values
44   X2 = df[["Rel_humidity"]].values
45   X = np.hstack([X1, X1*X2])   # Level & Level*Rel_humidity
46   y = df["Streamflow"].values
47   X_train, X_test, y_train, y_test = train_test_split(
48       X, y, test_size=0.33, random_state=11)
49
50   """
51   Linear Regression model fitting
52   """
53   nonlinearRegression = LinearRegression()
54   nonlinearRegression.fit(X_train, y_train)
55   b0, b1, b2 = nonlinearRegression.intercept_, \
56       nonlinearRegression.coef_[0], nonlinearRegression.coef_[1]
57   print("\n\nEstimated parameters: "
58       "\n    (b0, b1, b2) = ({:.2f}, {:.2f}, {:.2f})".
59       format(b0, b1, b2))
60   print("R-squared value: \n     R^2 = {:.2f}".
61       format(nonlinearRegression.score(X_train, y_train)))
62
63   """
64   Visualization of results
65   """
66   # Extract min max
67   xmin, xmax = np.min(X_train[:, 0]), np.max(X_train[:, 0])
68   ymin, ymax = np.min(X_train[:, 1]), np.max(X_train[:, 1])
69   zmin, zmax = np.min(y_train), np.max(y_train)
70
71   # Surface equation
72   XX = np.linspace(xmin, xmax, 20)
73   YY = np.linspace(ymin, ymax, 20)
74   xx, yy = np.meshgrid(XX, YY)
75   zz = b0 + b1 * xx + b2 * yy
76
77   # Surface plot and Scatter 3d plot of points
78   fig = plt.figure(figsize=(7, 6))
79   ax = plt.axes(projection='3d')
80   ax.plot_surface(xx, yy, zz,
81                   rstride=1, cstride=1, alpha=0.5)
82   ax.scatter(X_train[:, 0], X_train[:, 1], y_train,
83              marker='o', edgecolors='black', c='red', s=30)
84
85   # Tick numbers and locations
86   xtick_loc = np.linspace(xmin, xmax, 5)
87   ytick_loc = np.linspace(ymin, ymax, 5)
88   ztick_loc = np.linspace(zmin, zmax, 5)
89   ax.set_xticks(xtick_loc)
90   ax.set_yticks(ytick_loc)
91   ax.set_zticks(ztick_loc)
```

```
82  # Labels
83  ldist = 9
84  ax.set_xlabel('Press. (X)', labelpad=ldist)
85  ax.set_ylabel('Level * Rel_humidity (Y)', labelpad=ldist + 4)
86  ax.set_zlabel('Streamflow (Z)', labelpad=ldist)
87
88  # Tidying up
89  ax.view_init(elev=30, azim=-135)
90  ax.axis('auto')
91  plt.tight_layout()
92  plt.show()
93  plt.savefig("nonlinearReg_.pdf", dpi=300)
94
95  print("Done!!")
```

```
● ● ●          Terminal

Estimated parameters:
    (b0, b1, b2) = (-48144.80, 288.41, -0.02)
R-squared value:
    R^2 = 0.93
```

First, the program reads the Godavari River dataset from a CSV file using pandas. It then checks the "time" field to ensure it is in the correct datetime format. The data, consisting of fields such as "time", "Pressure", "Rel_humidity", "Level", and "Streamflow", is resampled to a monthly frequency using pandas' resample() function. This is done to reduce the frequency of the data points from daily or hourly to monthly, thereby downsampling the time series data.

Following this, the "Level" field is chosen as the independent variable (X), and its square root is also appended to it. This effectively transforms the original variable, allowing the model to capture nonlinear relationships. The "Streamflow" field is considered the dependent variable (y). The data is then split into training and test sets using the train_test_split() function from sklearn, with 33% of the data kept for testing.

A Linear Regression model from sklearn is then fit to the training data. Despite being a "linear" regression model, it is used here to fit a nonlinear relationship due to the transformation in the second independent variable. The fitted model's parameters, including the intercept and coefficients for "Level" and "Level×Rel_humidity", are printed, along with the R-squared value, which measures the proportion of variance in the dependent variable which can be predicted from the independent variable.

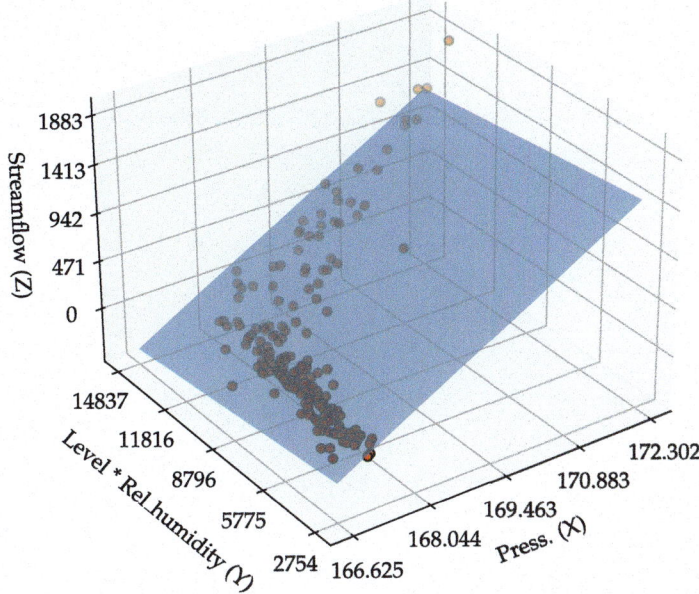

Fig. 6.3 Nonlinear regression of the Streamflow data on Level and (Level×Rel_humidity) data for the Godavari River

Finally, the program generates a three-dimensional plot (Fig. 6.3) to visualize the results. The "Level" and "Level × Rel_humidity" fields are plotted on the x and y-axes, respectively, while the "Streamflow" is on the z-axis. The latter variable is the nonlinear term of the regression. The plot also shows the estimated regression surface with the data points.

Chapter 7
Hydrological Time Series Analysis

Forecasting a time series into the future is one of the most common applications of statistical modeling in hydrology. Typically, hydrologists are interested in forecasting streamflow and floods, which itself is a complex nonlinear phenomenon controlled by meteorological and geomorphological factors. Apart from streamflow, hydrologists typically also attempt to forecast soil moisture, groundwater level, and water quality parameters. Various time series statistical methods can be applied to a long record of observations, in order to uncover trends, patterns, and cycles.

Knowledge of these temporal patterns is essential for two reasons—firstly, understanding the underlying mechanisms governing the relationships between different hydrological variables, and secondly, forecasting future hydrological conditions based on our understanding of past patterns. For example, one could use historical records of precipitation, temperature, and streamflow data for managing water resources, forecasting floods, and planning for droughts. In this era of climate change, such predictive models form a crucial input for assessing the potential impacts on water availability and adaptation strategies.

One such technique is Autoregressive Integrated Modeling Average or ARIMA, a standard method used to analyze hydrological time series. Their ability to capture seasonality, autocorrelation, and trends in time series data makes them suitable for diverse applications. For example, one can predict the monthly rainfall with an ARIMA model using past observations, while its model parameters would indicate the degree of seasonality or persistence.

Another area where time series analysis is increasingly important is the impact of human activities on hydrological systems. A comparison of time series data before and after a specific event or human intervention (like a change in land use or dam construction) can also reveal the impact on groundwater levels, river flows, or water quality.

A. Kumar and M. Saharia, *Python for Water and Environment*, Innovations in Sustainable Technologies and Computing, https://doi.org/10.1007/978-981-99-9408-3_7

Time series forecasting and analysis are essential tools in hydrology, enhancing our ability to predict future conditions and understand hydrological systems. In the following sections, we describe some commonly used techniques for forecasting time series and statistical tests, along with example codes in Python.

7.1 Stationarity, Trend, and Periodicity

Stationarity, trend, seasonality, and periodicity are fundamental concepts in time series analysis. Stationarity refers to the property of a time series where its statistical features such as mean, variance, and autocorrelation remain constant over time. This is crucial because many statistical modeling techniques assume or require the time series to be stationary.

Trend refers to the long-term movement in a time series. It can either be upward, downward, or horizontal, signifying an increase, decrease, or no change in the variable of interest over time, respectively.

Seasonality is the characteristic of a time series in which the data exhibits regular, predictable changes that recur every calendar season. This can be due to seasonal effects like weather changes or other cyclical factors affecting the variable of interest.

Periodicity, similar to seasonality, refers to patterns that repeat over predictable, fixed periods of time. The difference that lies in the periodic patterns is not tied to the calendar. For example, a certain pattern in a time series that repeats every 10 data points demonstrates periodicity. The identification of these components within a time series can help improve the accuracy of forecast models and lead to more insightful analyses.

```python
"""
Program to check for stationarity of a
time series signal and decompose it
into trend and seasonal components
"""

# Import libraries
import pandas as pd
import matplotlib.pyplot as plt
from statsmodels.tsa.stattools import adfuller as ADF
from statsmodels.tsa.seasonal import seasonal_decompose


"""
Load dataset
"""
data = pd.read_csv(
    filepath_or_buffer="../data/Godavari.csv",
    sep=",",
```

```
1        header=0).dropna()
2    print("\nChecking data:")
3    try:
4        data["time"] = pd.to_datetime(
5            data['time'], infer_datetime_format=True)
6        print("   Date format is okay!\n")
7    except ValueError:
8        print("   Encountered error!\n")
9        pass
10   df = data[["time", "Streamflow"]]
11   del data
12   print("Read data file")
13
14
15   """
16   Downsample the time series
17   """
18   resampled = df.resample('2M', on="time").mean()
19
20
21   """
22   Transform the data
23   """
24   def transform(x):
25       x = x - minx + 10.0
26       return x
27
28
29   def inverse_transform(x):
30       x = x - 10.0 + minx
31       return x
32
33
34   """
35   Transform data
36   """
37   minx = resampled.min()
38   resampled = transform(resampled)
39
40
41   """
42   Decompose a signal (multiplicative/additive)
43   """
44   decompose_result_mult = seasonal_decompose(
45       resampled, model="multiplicative")
46   fig, ax = plt.subplots(nrows=3, ncols=1, sharex=True,
47                          figsize=(8, 6))
```

```python
37  ax[0].plot(decompose_result_mult.observed, label='Time series')
38  ax[1].plot(decompose_result_mult.seasonal, label='Seasonal')
39  ax[2].plot(decompose_result_mult.trend, label='Trend')
40  ax[0].set_ylabel('Series')
41  ax[1].set_ylabel('Seasonal')
42  ax[2].set_ylabel('Trend')
43  ax[2].set_xlabel('Year')
44  ax[0].grid(ls='--')
45  ax[1].grid(ls='--')
46  ax[2].grid(ls='--')
47  plt.tight_layout()
48  plt.savefig('seasonal.pdf', dpi=300)
49
50
51  """
52  Stationarity check
53  """
54  def stationarity_adf_test(x, alpha=0.05):
55      adftest_res = ADF(x, autolag="AIC")
56      dfout = pd.Series(
57          adftest_res[0:4],
58          index=["ADF statistic", "ADF p-value",
59                 "ADF lags used", "ADF number of obs used"])
60      for key, value in adftest_res[4].items():
61          dfout["    Critical Value (%s)" % key] = value
62      print(dfout)
63      if dfout["ADF p-value"] > alpha:
64          print("    Result: Non-stationary time series", "\n")
65      else:
66          print("    Result: Stationary time series", "\n")
67
68
69  print("\nChecking stationarity:")
70  stationarity_adf_test(resampled)
```

```
●  ●  ●          Terminal

>>> Checking data:
>>>     Date format is okay!
>>>
>>> Read data file
>>>
>>> Checking stationarity:
>>> ADF statistic                -2.697121
>>> ADF p-value                   0.074538
>>> ADF lags used                12.000000
>>> ADF number of obs used      137.000000
>>>     Critical Value (1%)      -3.479007
>>>     Critical Value (5%)      -2.882878
>>>     Critical Value (10%)     -2.578149
>>> dtype: float64
>>>     Result: Non-stationary time series
```

The above program begins by importing the necessary libraries, including pandas for data handling, matplotlib for data visualization, and statsmodels for time series analysis.

We then load the dataset from a CSV file named "Godavari.csv". The CSV file is assumed to contain a time series of streamflow data with timestamps. The dataset is read into a pandas dataframe, and any missing values are dropped. The "time" column of the dataframe is converted to datetime format for proper time series analysis. If the datetime conversion is successful, a message is printed to the console confirming the correct date format.

The next part of the program down-samples the original time series data, taking the mean value for every 2-month period. This is done to simplify the time series and reduce the influence of short-term fluctuations.

The program then transforms the data to ensure all values are positive by subtracting the minimum value from all data points and adding 10. The transformation helps make all the values in the time series positive, making the forecast model more stable. This transformed dataset is stored in the variable "resampled".

The transformed, resampled time series is then decomposed into its trend and seasonal components using the "seasonal_decompose()" function from the statsmodels library. The seasonal_decompose function splits the time series into three parts: the original time series, the seasonal component (repeating patterns at specific intervals), and the trend component. We then plot the three parts on a shared x-axis for visual comparison.

Finally, the program checks the stationarity of the transformed, resampled time series data. Stationarity is a key assumption in many time series analysis methods. A time series is said to be stationary if its properties, such as mean and variance, do not change over time. The program uses the Augmented Dickey-Fuller (ADF) test, a common statistical test for stationarity, implemented in the statsmodels library. The results of the test are printed to the console, indicating whether the time series is stationary or non-stationary based on the p-value from the ADF test. A graphical display of the decomposition is shown in Fig. 7.1.

7.2 Common Forecasting Methods

7.2.1 Autoregression (AR)

Autoregressive (AR) time series forecasting is a statistical approach whereby future values are predicted based on a weighted sum of past observations. The premise is that past values in the series have an effect on what happens next. An AR model uses the direct relationship between an observation and a certain number of lagged observations (i.e., observations at previous points in time). The "order" of the AR model denotes how many previous observations are used. The adjoining code demonstrates a code to make a time series forecasting using AR.

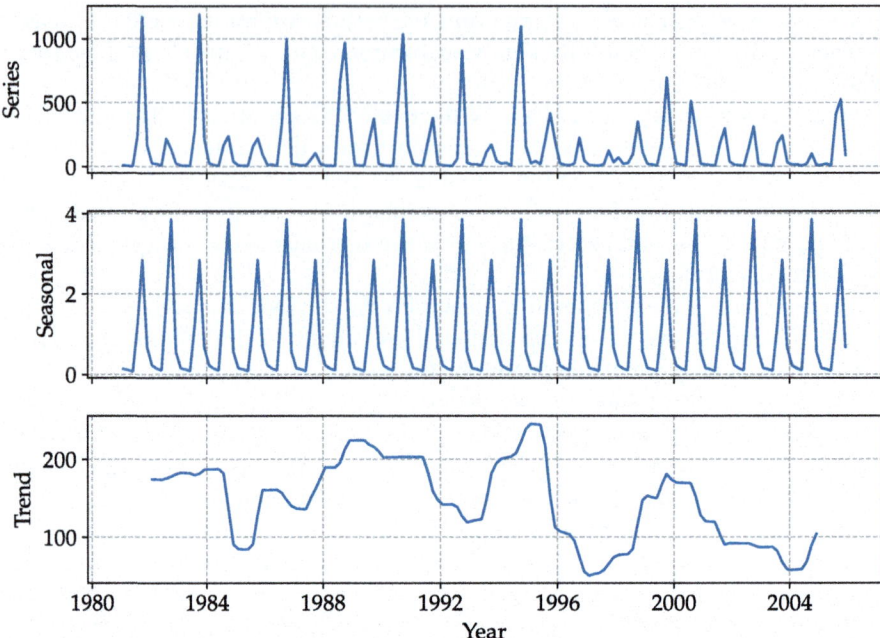

Fig. 7.1 Decomposition of the streamflow data of the Godavari River into its seasonal component and trend, assuming that the amplitude of the seasonal component increases or decreases with the trend. The option "multiplicative" is indicative of the same in the code. The trend is representative of the long-term movement/pattern in the time series data. The seasonal component shows the repeating patterns/fluctuations that occur repeatedly

```
1   """
2   Program to do time series modeling
3   using the Autoregression model
4   """
5
6   # Import libraries
7   import pandas as pd
8   import numpy as np
9   import matplotlib.pyplot as plt
10  from statsmodels.graphics.tsaplots import plot_acf, plot_pacf
11  from statsmodels.tsa.stattools import adfuller as ADF
12  from statsmodels.tsa.ar_model import AutoReg as AR
13  from sklearn.metrics import mean_squared_error as MSE
14  import seaborn as sns
15
16  savePlots = 1
```

```python
def plotGraph(df_, label, figsize, indicator, title,
              save=savePlots):
    if save:
        if indicator == 0: # Time series plot
            df_.plot(figsize=figsize)
            plt.xlabel(label[0])
            plt.ylabel(label[1])
            plt.legend()
        elif indicator == 1: # Histogram
            fig, ax = plt.subplots(2, 1, figsize=figsize,
                                   sharex=True)
            df_["Streamflow"].hist(ax=ax[0])
            df_["Streamflow"].plot(kind='kde', ax=ax[1])
            ax[0].set_title("")
            ax[0].grid(ls="--")
            ax[1].grid(ls="--")
            ax[1].set_xlabel(label[0])
            ax[1].set_ylabel(label[1])
        elif indicator == 2: # Boxplot
            plt.figure(figsize=figsize)
            sns.boxplot(x=df_["Streamflow"].index.year,
                        y=df_["Streamflow"])
        elif indicator == 3: # Autocorrelation plot
            fig, ax = plt.subplots(2, 1, figsize=figsize,
                                   sharex=True)
            plot_acf(x=df_["Streamflow"], lags=40, ax=ax[0])
            ax[0].set_title("")
            ax[0].set_ylabel("ACF")
            plot_pacf(x=df_["Streamflow"],
                      lags=40, method="ywm", ax=ax[1])
            ax[1].set_xlabel("lag")
            ax[1].set_ylabel("PACF")
        elif indicator == 4: # Prediction results
            fig, ax = plt.subplots()
            df_.plot(y="Test series",
                     use_index=True, style="-x",
                     lw=3, ms=8, ax=ax)
            df_.plot(y="Predicted series",
                     use_index=True, style="-o",
                     lw=3, ms=8, alpha=0.6,
                     ax=ax)
            ax.grid('on', ls="--", which='minor', axis='both')
            plt.xlabel(label[0])
            plt.ylabel(label[1])
        plt.title("")
```

```
44          plt.grid(ls="--")
45          plt.tight_layout()
46          plt.show()
47          plt.savefig(title + "_.pdf", dpi=300)
48
49
50  """
51  Load dataset
52  """
53  data = pd.read_csv(
54      filepath_or_buffer="../data/Godavari.csv",
55      sep=",",
56      header=0).dropna()
57  print("\nChecking data:")
58  try:
59      data["time"] = pd.to_datetime(
60          data['time'], infer_datetime_format=True)
61      print("   Date format is okay!\n")
62  except ValueError:
63      print("   Encountered error!\n")
64      pass
65  df = data[["time", "Streamflow"]]
66  del data
67  print("Read data file")
68
69
70  """
71  Resample:
72  Downsample the time series
73  """
74  resampled = df.resample('2M', on="time").mean()
75  plotGraph(df_=resampled,   # Time series plot
76            label=["Year", "Streamflow"],
77            figsize=(10, 4),
78            indicator=0,
79            title="ar-ts_resampled")
80
81  # Before transformation and standardization
82  plotGraph(df_=resampled,   # Histogram
83            label=["Streamflow", ""],
84            figsize=(7, 6),
85            indicator=1,
86            title="ar-histogram")
87  plotGraph(df_=resampled,   # Boxplot
88            label=["Year", "Streamflow"],
89            figsize=(14, 4),
```

```
85          indicator=2,
86          title="ar-boxplot")
87
88
89  """
90  Pre-processing:
91  Data transformation and Visualization
92  """
93
94
95  def transform_data(x):
96      x = np.log(x - xmin + 100.)
97      x = (x - meanx_) / stdx_
98      return x
99
100
101 def inverse_transform(x):
102     x = x * stdx_ + meanx_    # non-standardized
103     x = np.exp(x) + xmin - 100.   # inverse log positive
104     return x
105
106
107 xmin = np.array(resampled.min())[0]
108 meanx_ = np.array(resampled.mean())[0]
109 stdx_ = np.array(resampled.std())[0]
110
111 resampled = resampled.apply(transform_data)
112 print("\nDone data transformation")
113
114
115
116 # After transformation and standardization
117 plotGraph(df_=resampled,    # Time series plot
118           label=["Year", "$log($" + "Streamflow" + "$)$"],
119           figsize=(10, 4),
120           indicator=0,
121           title="ar-ts_resampled_after")
122 plotGraph(df_=resampled,    # Histogram
123           label=["Streamflow", ""],
124           figsize=(7, 6),
125           indicator=1,
126           title="ar-histogram_after")
127 plotGraph(df_=resampled,    # Boxplot
128           label=["Year", "Streamflow"],
129           figsize=(14, 4),
130           indicator=2,
131           title="ar-boxplot_after")
```

```
128    """
129    Stationarity check
130    """
131    def stationarity_adf_test(x, alpha=0.05):
132        adftest_res = ADF(x, autolag="AIC")
133        dfout = pd.Series(
134            adftest_res[0:4],
135            index=["ADF statistic", "ADF p-value",
136                    "ADF lags used", "ADF number of obs used"])
137        for key, value in adftest_res[4].items():
138            dfout["   Critical Value (%s)" % key] = value
139        print(dfout)
140        if dfout["ADF p-value"] > alpha:
141            print("   Result: Non-stationary time series", "\n")
142        else:
143            print("   Result: Stationary time series", "\n")
144
145
146    print("\nChecking stationarity:")
147    stationarity_adf_test(resampled)
148
149
150    """
151    Autocorrelation and Partial autocorrelation
152    """
153    plotGraph(df_=resampled,    # Autocorrelation plot
154            label=["", ""],
155            figsize=(7, 6),
156            indicator=3,
157            title="ar-acorr")
158
159
160    """
161    Data partition and Simple Autoregression fitting
162    """
163    trainmask = (resampled.index >= "1981-01-01") & \
164                (resampled.index <= "2002-12-31")
165    testmask = (resampled.index > "2002-12-31")
166    training_set = list(resampled["Streamflow"].loc[trainmask])
167    testing_set = list(resampled["Streamflow"].loc[testmask])
168    print("Created training and testing datasets\n")
169
170    """
171    Single-step predictions
172    using autoregression
173    """
```

```
170  forecasts = list()
171  for i in range(len(testing_set)):
172      modelAR = AR(training_set, lags=3).fit()
173      pred = np.squeeze(modelAR.forecast())
174      forecasts.append(pred)
175      training_set.append(testing_set[i])
176      print("Obs={:0.04f}, Pred={:0.04f}   {}/{}".format(
177          np.squeeze(inverse_transform(testing_set[i])),
178          np.squeeze(inverse_transform(pred)),
179          i + 1,
180          len(testing_set))
181      )
182  print("\nRoot mean square error:")
183  print("   Testset RMSE={:0.04}".
184      format(MSE(testing_set, forecasts)))
185
186  """
187  Result plots
188  """
189  result = {"Year": resampled["Streamflow"].loc[testmask].index,
190          "Test series":
191              inverse_transform(np.array(testing_set)),
192          "Predicted series":
193              inverse_transform(np.array(forecasts))}
194  df_res = pd.DataFrame.from_dict(data=result)
195  df_res.set_index("Year", inplace=True)
196  plotGraph(df_=df_res,  # Prediction results
197          label=["Year", "Streamflow"],
198          figsize=(10, 4),
199          indicator=4,
200          title="ar-result")
201  print("Done!")
```

The provided Python script performs time series analysis using the Autoregression (AR) model. The script is split into various sections, each performing a specific task in the process of analyzing the time series data.

In the beginning, it imports all the necessary libraries and defines a function called "plotGraph()". This function is created to generate and save different types of plots based on the indicator passed on to it, such as 0—for time series plot, 1—for histogram, 2—boxplot, 3—plot of autocorrelation, and 4—plot predictions. It helps in visualizing the time series data at various stages of preprocessing and modeling.

The script then loads a dataset from a CSV file named "Godavari.csv", ensuring that the data is free of null values. It attempts to convert the "time" column into a datetime format, to facilitate time series analysis. After the successful conversion, the data is resampled bi-monthly and visualized in different ways, including a time series plot, histogram, and boxplot.

In the pre-processing stage, the data is transformed to be suitable for further analysis. Data transformation consists of applying a logarithmic function and data

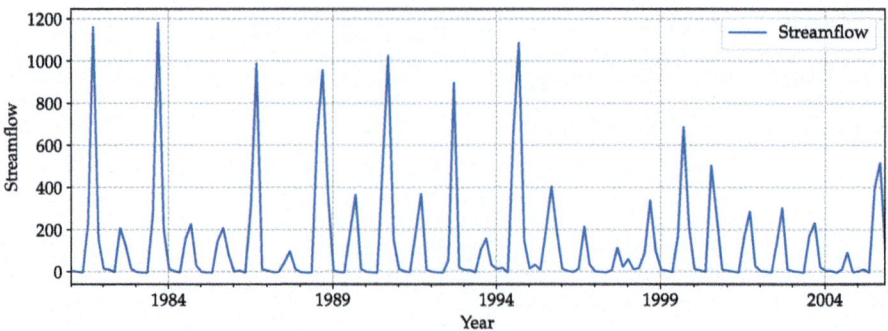

Fig. 7.2 Observed time series of the stream flow data of the Godavari River after resampling

standardization process; afterward, the data is visualized once more. The resampled time series plot is shown in Fig. 7.2, histogram in Fig. 7.3, and boxplot in Fig. 7.4. We only show these plots for the AR case and avoid redundant plots by not reshowing them in the upcoming time series analysis programs.

Next, the transformed data is subjected to an Augmented Dickey-Fuller (ADF) test to check stationarity. Crucial for modeling time series, a stationary time series is one where the properties do not depend on the time the series is observed. An

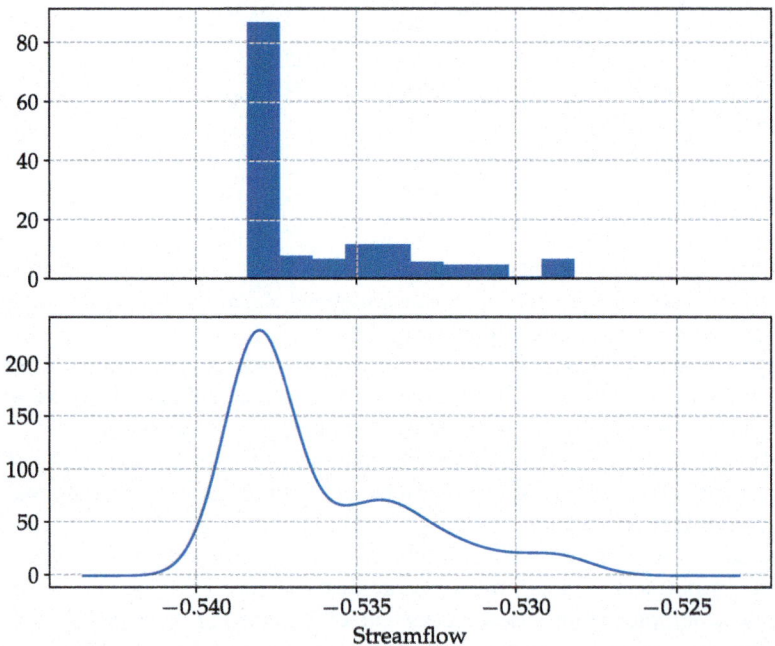

Fig. 7.3 Histogram of the observed stream flow time series data of the Godavari River after transformation

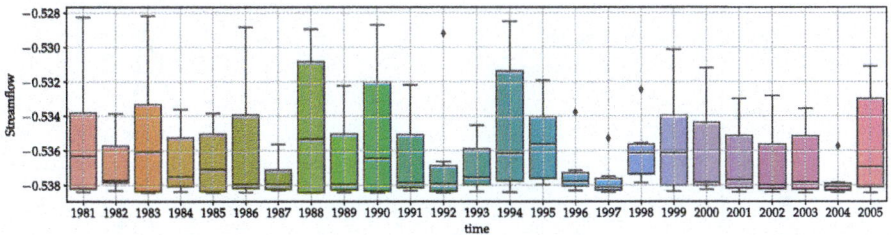

Fig. 7.4 Year-wise box plot of the observed time series data of the Godavari River after transformation

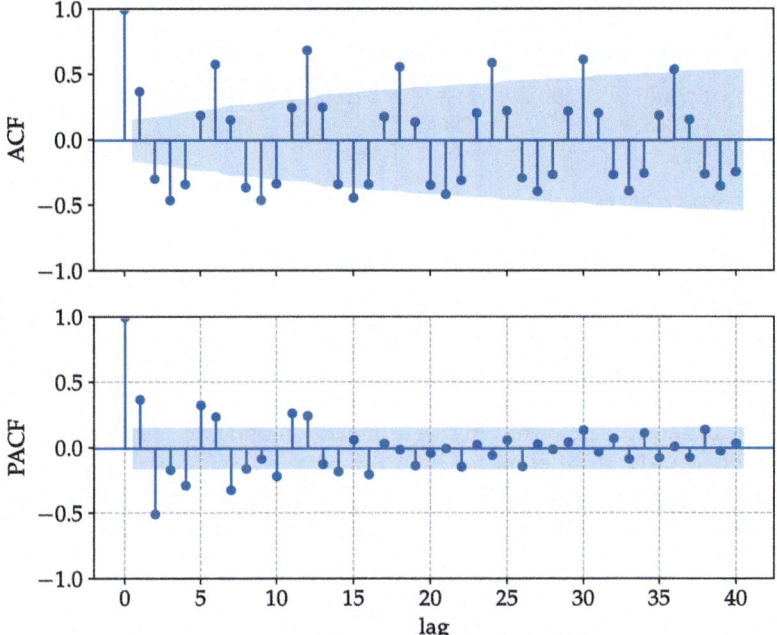

Fig. 7.5 Plot showing the autocorrelation and partial autocorrelation function computed on the observed data of the Godavari River after transformation

autocorrelation plot (Fig. 7.5) is also generated to visualize the time series' correlation with its own past and future values.

Finally, the data is partitioned into sets of training and testing. The autoregression model is then fit to the training data and used to predict values in the test data in a single-step fashion. This means each step in the test set is predicted one at a time, using the predicted values from the previous steps. The prediction result is shown in Fig. 7.6.

The root mean square error (RMSE) of the prediction on the test set is calculated to quantify the prediction error. Finally, the true values and predicted values in the test set are plotted for visual comparison. This can help the user evaluate the effectiveness of the autoregression model in predicting the time series data.

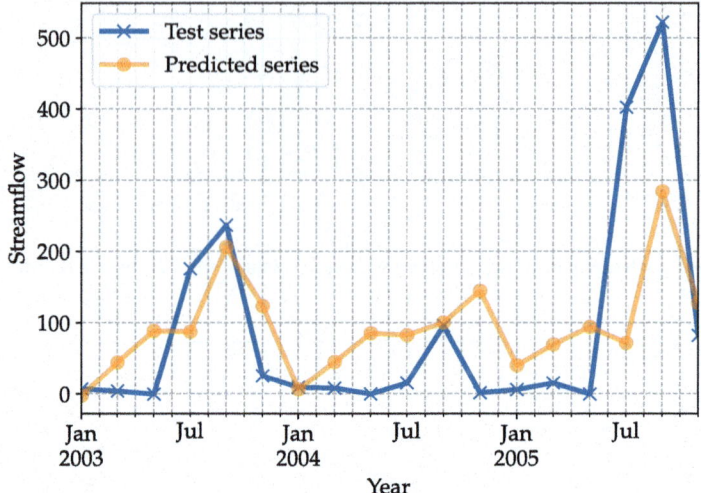

Fig. 7.6 Forecast result using the Autoregression forecast model

7.2.2 *Autoregressive Moving Average (ARMA)*

Autoregressive Moving Average (ARMA) combines two components: an Autoregressive (AR) part and a Moving Average (MA) part. The AR part analyzes the correlation between an observation and a certain number of lagged observations. The MA part models the error term as a linear combination of error terms occurring simultaneously at different times in the past. By using both past values and past forecast errors to model future values, ARMA can effectively capture complex autocorrelations in time series data, thus, improving the accuracy of predictions.

```
1   """
2   Program to do time series modeling
3   using the ARMA model
4   """
5
6   # Import libraries
7   import pandas as pd
8   import numpy as np
9   import matplotlib.pyplot as plt
10  from statsmodels.graphics.tsaplots import plot_acf, plot_pacf
11  from statsmodels.tsa.stattools import adfuller as ADF
12  from statsmodels.tsa.arima.model import ARIMA
13  from sklearn.metrics import mean_squared_error as MSE
14  import seaborn as sns
```

```
1   savePlots = 1
2
3
4   def plotGraph(df_, label, figsize, indicator, title,
5                 save=savePlots):
6       if save:
7           if indicator == 0: # Time series plot
8               df_.plot(figsize=figsize)
9               plt.xlabel(label[0])
10              plt.ylabel(label[1])
11              plt.legend()
12          elif indicator == 1: # Histogram
13              fig, ax = plt.subplots(2, 1, figsize=figsize,
14                                     sharex=True)
15              df_["Streamflow"].hist(ax=ax[0])
16              df_["Streamflow"].plot(kind='kde', ax=ax[1])
17              ax[0].set_title("")
18              ax[0].grid(ls="--")
19              ax[1].grid(ls="--")
20              ax[1].set_xlabel(label[0])
21              ax[1].set_ylabel(label[1])
22          elif indicator == 2: # Boxplot
23              plt.figure(figsize=figsize)
24              sns.boxplot(x=df_["Streamflow"].index.year,
25                          y=df_["Streamflow"])
26          elif indicator == 3: # Autocorrelation
27              fig, ax = plt.subplots(2, 1, figsize=figsize,
28                                     sharex=True)
29              plot_acf(x=df_["Streamflow"], lags=40, ax=ax[0])
30              ax[0].set_title("")
31              ax[0].set_ylabel("ACF")
32              plot_pacf(x=df_["Streamflow"],
33                        lags=40, method="ywm", ax=ax[1])
34              ax[1].set_xlabel("lag")
35              ax[1].set_ylabel("PACF")
36          elif indicator == 4: # Predicted result
37              fig, ax = plt.subplots()
38              df_.plot(y="Test series",
39                       use_index=True, style="-x",
40                       lw=3, ms=8, ax=ax)
41              df_.plot(y="Predicted series",
42                       use_index=True, style="-o",
43                       lw=3, ms=8, alpha=0.6,
44                       ax=ax)
45              ax.grid('on', ls="--", which='minor', axis='both')
46              plt.xlabel(label[0])
47              plt.ylabel(label[1])
```

```python
44          plt.title("")
45          plt.grid(ls="--")
46          plt.tight_layout()
47          plt.show()
48          plt.savefig(title + "_.pdf", dpi=300)
49
50
51  """
52  Load dataset
53  """
54  data = pd.read_csv(
55      filepath_or_buffer="../data/Godavari.csv",
56      sep=",",
57      header=0).dropna()
58  print("\nChecking data:")
59  try:
60      data["time"] = pd.to_datetime(
61          data['time'], infer_datetime_format=True)
62      print("   Date format is okay!\n")
63  except ValueError:
64      print("   Encountered error!\n")
65      pass
66  df = data[["time", "Streamflow"]]
67  del data
68  print("Read data file")
69
70
71  """
72  Resample:
73  Downsample the time series
74  """
75  resampled = df.resample('2M', on="time").mean()
76  plotGraph(df_=resampled,   # Time series plot
77           label=["Year", "Streamflow"],
78           figsize=(10, 4),
79           indicator=0,
80           title="arma-ts_resampled")
81
82  # Before transformation and standardization
83  plotGraph(df_=resampled,   # Histogram
84           label=["Streamflow", ""],
85           figsize=(7, 6),
86           indicator=1,
87           title="arma-histogram")
```

```python
87   plotGraph(df_=resampled,   # Boxplot
88            label=["Year", "Streamflow"],
89            figsize=(14, 4),
90            indicator=2,
91            title="arma-boxplot")
92
93
94   """
95   Pre-processing:
96   Data transformation and Visualization
97   """
98
99
100  def transform_data(x):
101      x = np.log(x - xmin + 100.)
102      x = (x - meanx_) / stdx_
103      return x
104
105
106  def inverse_transform(x):
107      x = x * stdx_ + meanx_   # non-standardized
108      x = np.exp(x) + xmin - 100.   # inverse log positive
109      return x
110
111
112  xmin = np.array(resampled.min())[0]
113  meanx_ = np.array(resampled.mean())[0]
114  stdx_ = np.array(resampled.std())[0]
115
116  resampled = resampled.apply(transform_data)
117  print("\nDone data transformation")
118
119
120
121  # After transformation and standardization
122  plotGraph(df_=resampled,   # Time series plot
123            label=["Year", "$log($" + "Streamflow" + "$)$"],
124            figsize=(10, 4),
125            indicator=0,
126            title="arma-ts_resampled_after")
127  plotGraph(df_=resampled,   # Histogram
128            label=["Streamflow", ""],
129            figsize=(7, 6),
130            indicator=1,
131            title="arma-histogram_after")
```

```
122  plotGraph(df_=resampled,   # Boxplot
123            label=["Year", "Streamflow"],
124            figsize=(14, 4),
125            indicator=2,
126            title="arma-boxplot_after")
127
128
129  """
130  Stationarity check
131  """
132  def stationarity_adf_test(x, alpha=0.05):
133      adftest_res = ADF(x, autolag="AIC")
134      dfout = pd.Series(
135          adftest_res[0:4],
136          index=["ADF statistic", "ADF p-value",
137                 "ADF lags used", "ADF number of obs used"])
138      for key, value in adftest_res[4].items():
139          dfout["   Critical Value (%s)" % key] = value
140      print(dfout)
141      if dfout["ADF p-value"] > alpha:
142          print("   Result: Non-stationary time series", "\n")
143      else:
144          print("   Result: Stationary time series", "\n")
145
146
147  print("\nChecking stationarity:")
148  stationarity_adf_test(resampled)
149
150
151  """
152  Autocorrelation and Partial autocorrelation
153  """
154  plotGraph(df_=resampled,   # Autocorrelation plot
155            label=["", ""],
156            figsize=(7, 6),
157            indicator=3,
158            title="arma-acorr")
159
160
161  """
162  Data partition and ARIMA fitting
163  """
164  trainmask = (resampled.index >= "1981-01-01") & \
165              (resampled.index <= "2002-12-31")
166  testmask = (resampled.index > "2002-12-31")
167  training_set = list(resampled["Streamflow"].loc[trainmask])
168  testing_set = list(resampled["Streamflow"].loc[testmask])
```

```
199  print("Created training and testing datasets\n")
200
201  """
202  Single-step predictions
203  order = (p,q,d)
204  order = (lags, difference, moving average)
205  """
206  forecasts = list()
207  for i in range(len(testing_set)):
208      modelARIMA = ARIMA(training_set, order=(5, 0, 0)).fit()
209      pred = np.squeeze(modelARIMA.forecast())
210      forecasts.append(pred)
211      training_set.append(testing_set[i])
212      print("Obs={:0.04f}, Pred={:0.04f}     {}/{}".format(
213          np.squeeze(inverse_transform(testing_set[i])),
214          np.squeeze(inverse_transform(pred)),
215          i + 1,
216          len(testing_set))
217      )
218  print("\nRoot mean square error:")
219  print("    Testset RMSE={:0.04}".
220      format(MSE(testing_set, forecasts)))
221
222  """
223  Result plots
224  """
225  result = {"Year": resampled["Streamflow"].loc[testmask].index,
226           "Test series":
227                  inverse_transform(np.array(testing_set)),
228           "Predicted series":
229                  inverse_transform(np.array(forecasts))}
230  df_res = pd.DataFrame.from_dict(data=result)
231  df_res.set_index("Year", inplace=True)
232  plotGraph(df_=df_res,   # Prediction results
233          label=["Year", "Streamflow"],
234          figsize=(10, 4),
235          indicator=4,
236          title="arma-result")
237  print("Done!")
```

This program performs time series modeling using the Autoregressive Moving Average (ARMA) model, specifically applied to a dataset of streamflow measurements from the Godavari River. The program uses several Python libraries, such as pandas, NumPy, matplotlib, statsmodels, and sklearn, all imported to assist with the data analysis and modeling.

The script begins by defining a function "plotGraph()" for creating and saving various types of plots. This function is built to handle time series, histogram, box plot, and autocorrelation plot, among others. The various indicators (0—time series plot, 1—histogram, 2—boxplot, 3—autocorrelation plot, 4—predicted result) define the type of plot to be generated.

Following this, the script proceeds to load the dataset using pandas' "read_csv()" function. It ensures the date format is correctly interpreted, extracts the relevant columns, and prints a confirmation message.

Several plots are generated to inspect the data that has been resampled visually. The time series frequency, originally at its initial resolution, is then reduced to a bi-monthly one.

Next, the program transforms the data, applying a logarithmic function and standardizing it, i.e., the mean is subtracted and then divided by the standard deviation. This transformation is crucial to stabilize variance and achieve a more Gaussian-like distribution. Both the transformed and original data are plotted for comparison.

The stationarity of the transformed data is tested using the Augmented Dickey-Fuller test, implemented via the "stationarity_adf_test()" function. The null hypothesis of the test is that a unit root is present in the time series sample. If the p-value obtained from the test is greater than the specified significance level (0.05), the script concludes that the time series is non-stationary.

The script then proceeds to analyze the autocorrelation and partial autocorrelation of the data, providing visual plots to help identify potential AR and MA components for the ARMA model.

The data is split into a training set, containing data from 1981 to 2002, and a testing set, containing data from 2003 onward. The model is then fitted iteratively on the expanding window of the training set, making single-step predictions at each step. The prediction result is shown in Fig. 7.7.

Lastly, the prediction results are displayed in terms of root mean square error (RMSE) for the test set, and a line plot is generated to compare the predicted series

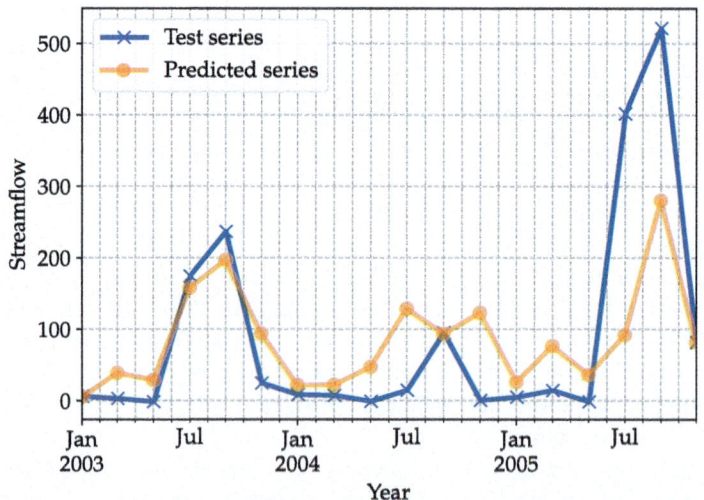

Fig. 7.7 Forecast result due to Autoregressive Moving Average model

with the actual test series. The RMSE serves as a measure of the differences between the values predicted by the model and the actual values. The line plot is useful to visually compare how closely the predictions align with the actual values.

This program showcases a thorough application of the ARMA model for time series data analysis, from data loading and preprocessing, to model fitting and result visualization.

7.2.3 Autoregressive Integrated Moving Average (ARIMA)

The Autoregressive Integrated Moving Average (ARIMA) model combines autoregression, differencing, and moving averages into a single approach to predict future data points. The autoregressive part analyzes the correlation between an observation and a certain number of lagged observations. The integrated part involves differencing the data to remove the trend and make the series stationary. The moving average part models the error term as a combination of past error terms. By tuning these components appropriately, ARIMA can capture complex temporal structures for forecasting.

```
1   """
2   Program to do time series modeling
3   using the ARIMA model
4   """
5
6   # Import libraries
7   import pandas as pd
8   import numpy as np
9   import matplotlib.pyplot as plt
10  from statsmodels.graphics.tsaplots import plot_acf
11  from statsmodels.graphics.tsaplots import plot_pacf
12  from statsmodels.tsa.stattools import adfuller as ADF
13  from statsmodels.tsa.arima.model import ARIMA
14  from sklearn.metrics import mean_squared_error as MSE
15  import seaborn as sns
16
17  # Save plots??
18  savePlots = 1
19
20
21  def plotGraph(df_, label, figsize, indicator, title,
22                  save=savePlots):
23      if save:
24          if indicator == 0: # time series plot
25              df_.plot(figsize=figsize)
26              plt.xlabel(label[0])
```

```
plt.ylabel(label[1])
plt.legend()
elif indicator == 1: # histogram
    fig, ax = plt.subplots(2, 1, figsize=figsize,
                           sharex=True)
    df_["Streamflow"].hist(ax=ax[0])
    df_["Streamflow"].plot(kind='kde', ax=ax[1])
    ax[0].set_title("")
    ax[0].grid(ls="--")
    ax[1].grid(ls="--")
    ax[1].set_xlabel(label[0])
    ax[1].set_ylabel(label[1])
elif indicator == 2: # boxplot
    plt.figure(figsize=figsize)
    sns.boxplot(x=df_["Streamflow"].index.year,
                y=df_["Streamflow"])
elif indicator == 3: # autocorrelation
    fig, ax = plt.subplots(2, 1, figsize=figsize,
                           sharex=True)
    plot_acf(x=df_["Streamflow"], lags=40, ax=ax[0])
    ax[0].set_title("")
    ax[0].set_ylabel("ACF")
    plot_pacf(x=df_["Streamflow"],
              lags=40, method="ywm", ax=ax[1])
    ax[1].set_xlabel("lag")
    ax[1].set_ylabel("PACF")
elif indicator == 4: # predicted result
    fig, ax = plt.subplots()
    df_.plot(y="Test series",
             use_index=True, style="-x",
             lw=3, ms=8, ax=ax)
    df_.plot(y="Predicted series",
             use_index=True, style="-o",
             lw=3, ms=8, alpha=0.6,
             ax=ax)
    ax.grid('on', ls="--", which='minor', axis='both')
    plt.xlabel(label[0])
    plt.ylabel(label[1])
plt.title("")
plt.grid(ls="--")
plt.tight_layout()
plt.show()
plt.savefig(title + "_.pdf", dpi=300)


"""
Load dataset
"""
```

```
43  data = pd.read_csv(
44      filepath_or_buffer="../data/Godavari.csv",
45      sep=",",
46      header=0).dropna()
47  print("\nChecking data:")
48  try:
49      data["time"] = pd.to_datetime(
50          data['time'], infer_datetime_format=True)
51      print("   Date format is okay!\n")
52  except ValueError:
53      print("   Encountered error!\n")
54      pass
55  df = data[["time", "Streamflow"]]
56  del data
57  print("Read data file")
58
59
60  """
61  Resample:
62  Downsample the time series
63  """
64  resampled = df.resample('2M', on="time").mean()
65  plotGraph(df_=resampled,   # Time series plot
66            label=["Year", "Streamflow"],
67            figsize=(10, 4),
68            indicator=0,
69            title="arima-ts_resampled")
70
71  # Before transformation and standardization
72  plotGraph(df_=resampled,   # Histogram
73            label=["Streamflow", ""],
74            figsize=(7, 6),
75            indicator=1,
76            title="arima-histogram")
77  plotGraph(df_=resampled,   # Boxplot
78            label=["Year", "Streamflow"],
79            figsize=(14, 4),
80            indicator=2,
81            title="arima-boxplot")
82
83
84  """
85  Pre-processing:
86  Data transformation and Visualization
87  """
88
89
90  def transform_data(x):
91      x = np.log(x - xmin + 100.)
```

```
124        x = (x - meanx_) / stdx_
125        return x
126
127
128    def inverse_transform(x):
129        x = x * stdx_ + meanx_   # non-standardized
130        x = np.exp(x) + xmin - 100.   # inverse log positive
131        return x
132
133
134    xmin = np.array(resampled.min())[0]
135    meanx_ = np.array(resampled.mean())[0]
136    stdx_ = np.array(resampled.std())[0]
137
138    resampled = resampled.apply(transform_data)
139    print("\nDone data transformation")
140
141
142
143    # After transformation and standardization
144    plotGraph(df_=resampled,   # Time series plot
145            label=["Year", "$log($" + "Streamflow" + "$)$"],
146            figsize=(10, 4),
147            indicator=0,
148            title="arima-ts_resampled_after")
149    plotGraph(df_=resampled,   # Histogram
150            label=["Streamflow", ""],
151            figsize=(7, 6),
152            indicator=1,
153            title="arima-histogram_after")
154    plotGraph(df_=resampled,   # Boxplot
155            label=["Year", "Streamflow"],
156            figsize=(14, 4),
157            indicator=2,
158            title="arima-boxplot_after")
159
160
161    """
162    Stationarity check
163    """
164    def stationarity_adf_test(x, alpha=0.05):
165        adftest_res = ADF(x, autolag="AIC")
166        dfout = pd.Series(
167            adftest_res[0:4],
168            index=["ADF statistic", "ADF p-value",
169                   "ADF lags used", "ADF number of obs used"])
```

```
164    for key, value in adftest_res[4].items():
165        dfout["    Critical Value (%s)" % key] = value
166    print(dfout)
167    if dfout["ADF p-value"] > alpha:
168        print("    Result: Non-stationary time series", "\n")
169    else:
170        print("    Result: Stationary time series", "\n")
171
172
173 print("\nChecking stationarity:")
174 stationarity_adf_test(resampled)
175
176
177 """
178 Autocorrelation and Partial autocorrelation
179 """
180 plotGraph(df_=resampled,    # Autocorrelation plot
181           label=["", ""],
182           figsize=(7, 6),
183           indicator=3,
184           title="arima-acorr")
185
186
187 """
188 Data partition and ARIMA fitting
189 """
190 trainmask = (resampled.index >= "1981-01-01") & \
191             (resampled.index <= "2003-12-31")
192 testmask = (resampled.index > "2003-12-31")
193 training_set = list(resampled["Streamflow"].loc[trainmask])
194 testing_set = list(resampled["Streamflow"].loc[testmask])
195 print("Created training and testing datasets\n")
196
197 """
198 Single-step predictions
199 order = (p,q,d)
200 order = (lags, difference, moving average)
201 """
202 forecasts = list()
203 for i in range(len(testing_set)):
204     modelARIMA = ARIMA(training_set, order=(5, 0, 0)).fit()
205     pred = np.squeeze(modelARIMA.forecast())
206     forecasts.append(pred)
207     training_set.append(testing_set[i])
208     print("Obs={:0.04f}, Pred={:0.04f}    {}/{}".format(
209         np.squeeze(inverse_transform(testing_set[i])),
```

```
201              np.squeeze(inverse_transform(pred)),
202              i + 1,
203              len(testing_set))
204        )
205   print("\nRoot mean square error:")
206   print("   Testset RMSE={:0.04}".
207         format(MSE(testing_set, forecasts)))
208
209   """
210   Result plots
211   """
212   result = {"Year": resampled["Streamflow"].loc[testmask].index,
213             "Test series":
214                   inverse_transform(np.array(testing_set)),
215             "Predicted series":
216                   inverse_transform(np.array(forecasts))}
217   df_res = pd.DataFrame.from_dict(data=result)
218   df_res.set_index("Year", inplace=True)
219   plotGraph(df_=df_res,   # Prediction results
220             label=["Year", "Streamflow"],
221             figsize=(10, 4),
222             indicator=4,
223             title="arima-result")
224   print("Done!")
```

This program carries out time series modeling using the ARIMA model, specifically designed for the analysis of the Godavari River streamflow. The first part of the program sets up necessary packages and defines a general function "plotGraph()" for creating different types of plots throughout the analysis. The function accepts different types of indicators to generate various plots such as time series plot (0), histogram (1), boxplot (2), autocorrelation plot (3), and prediction results (4).

The data loading step loads a .csv data file, containing the streamflow data of the Godavari River, into a pandas dataframe. Any missing values are removed to create a clean dataset. In addition, the code checks if the time column in the dataset is in the correct datetime format.

The loaded data is then downsampled or resampled to a lower frequency ("2M" or 2-month frequency) to reduce data volume and complexity. The resampled data is then visualized through a series of plots including time series plot, histogram, and boxplot.

The next step in the process is data transformation. This includes standardizing the data by subtracting the mean and dividing it by the standard deviation, and then taking the log transformation of the data to make it suitable for further analysis. The same plots are generated following transformation and standardization to visualize the transformed data.

The program performs an Augmented Dickey-Fuller (ADF) test for stationarity in time series data. Stationarity is an essential assumption in time series analysis.

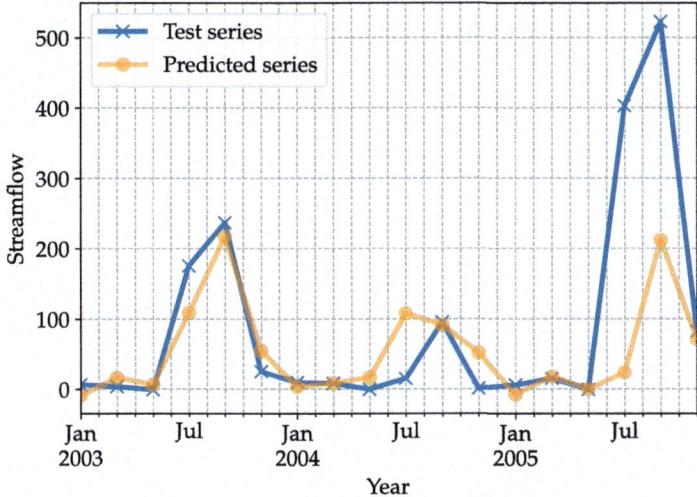

Fig. 7.8 Forecast result using the Autoregressive Integrated Moving Average model

After confirming stationarity, autocorrelation, and partial autocorrelation plots are generated to observe the correlation between the data points and their lags.

Next, the resampled and transformed data is partitioned into a training set (from "1981-01-01" to "2003-12-31") and a testing set (after "2003-12-31"). The ARIMA model is fitted to the training data, and single-step predictions are made. These predictions are made in a loop, each time adding the new test data point to the training set for the next prediction. The prediction result is shown in Fig. 7.8.

Finally, the program computes and prints the root mean square error (RMSE) of the test set, which quantifies the discrepancy between the model's predictions and the actual values. It also generates a plot to visually compare the test and predicted series.

7.2.4 Simple Exponential Smoothing (SES)

Simple Exponential Smoothing (SES) is a time series forecasting method for univariate data that does not consider trend and seasonality. It uses a weighted average of past observations as the forecast, where the weights decrease exponentially as observations come from further in the past—the smallest weights being associated with the oldest observations. The rate at which the weights decrease is a parameter of the method, referred to as the "smoothing constant". This parameter is chosen to minimize a measure of forecast error, such as Mean Squared Error. SES is particularly useful for data with no clear trend or seasonal pattern (Fig. 7.9).

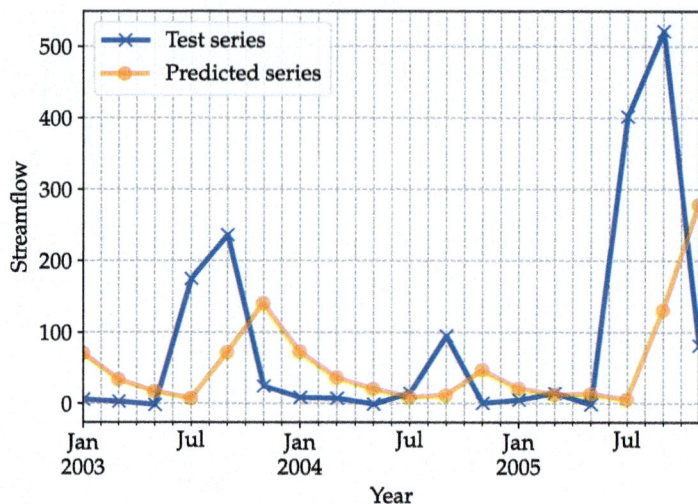

Fig. 7.9   Forecast results using the simple exponential smoothing model

```
1    """
2    Program to do time series modeling
3    using the Simple exponential smoothing
4    model
5    """
6
7    # Import libraries
8    import pandas as pd
9    import numpy as np
10   import matplotlib.pyplot as plt
11   from statsmodels.graphics.tsaplots import plot_acf, plot_pacf
12   from statsmodels.tsa.stattools import adfuller as ADF
13   from statsmodels.tsa.api import SimpleExpSmoothing
14   from sklearn.metrics import mean_squared_error as MSE
15   import seaborn as sns
16
17   savePlots = 1
18
19
20   def plotGraph(df_, label, figsize, indicator, title,
21                 save=savePlots):
22       if save:
23           if indicator == 0: # time series plot
24               df_.plot(figsize=figsize)
25               plt.xlabel(label[0])
26               plt.ylabel(label[1])
27               plt.legend()
```

```
elif indicator == 1: # histogram
    fig, ax = plt.subplots(2, 1, figsize=figsize,
                           sharex=True)
    df_["Streamflow"].hist(ax=ax[0])
    df_["Streamflow"].plot(kind='kde', ax=ax[1])
    ax[0].set_title("")
    ax[0].grid(ls="--")
    ax[1].grid(ls="--")
    ax[1].set_xlabel(label[0])
    ax[1].set_ylabel(label[1])
elif indicator == 2: # boxplot
    plt.figure(figsize=figsize)
    sns.boxplot(x=df_["Streamflow"].index.year,
                y=df_["Streamflow"])
elif indicator == 3: # autocorrelation
    fig, ax = plt.subplots(2, 1, figsize=figsize,
                           sharex=True)
    plot_acf(x=df_["Streamflow"], lags=40, ax=ax[0])
    ax[0].set_title("")
    ax[0].set_ylabel("ACF")
    plot_pacf(x=df_["Streamflow"],
              lags=40, method="ywm", ax=ax[1])
    ax[1].set_xlabel("lag")
    ax[1].set_ylabel("PACF")
elif indicator == 4: # predicted result
    fig, ax = plt.subplots()
    df_.plot(y="Test series",
             use_index=True, style="-x",
             lw=3, ms=8, ax=ax)
    df_.plot(y="Predicted series",
             use_index=True, style="-o",
             lw=3, ms=8, alpha=0.6,
             ax=ax)
    ax.grid('on', ls="--", which='minor', axis='both')
    plt.xlabel(label[0])
    plt.ylabel(label[1])
plt.title("")
plt.grid(ls="--")
plt.tight_layout()
plt.show()
plt.savefig(title + "_.pdf", dpi=300)


"""
Load dataset
"""
```

```
45  data = pd.read_csv(
46      filepath_or_buffer="../data/Godavari.csv",
47      sep=",",
48      header=0).dropna()
49  print("\nChecking data:")
50  try:
51      data["time"] = pd.to_datetime(
52          data['time'], infer_datetime_format=True)
53      print("   Date format is okay!\n")
54  except ValueError:
55      print("   Encountered error!\n")
56      pass
57  df = data[["time", "Streamflow"]]
58  del data
59  print("Read data file")
60
61
62  """
63  Resample:
64  Downsample the time series
65  """
66  resampled = df.resample('2M', on="time").mean()
67  plotGraph(df_=resampled,  # Time series plot
68            label=["Year", "Streamflow"],
69            figsize=(10, 4),
70            indicator=0,
71            title="ses-ts_resampled")
72
73  # Before transformation and standardization
74  plotGraph(df_=resampled,  # Histogram
75            label=["Streamflow", ""],
76            figsize=(7, 6),
77            indicator=1,
78            title="ses-histogram")
79  plotGraph(df_=resampled,  # Boxplot
80            label=["Year", "Streamflow"],
81            figsize=(14, 4),
82            indicator=2,
83            title="ses-boxplot")
84
85
86  """
87  Pre-processing:
88  Data transformation and Visualization
89  """
```

```python
87  def transform_data(x):
88      x = np.log(x - xmin + 100.)
89      x = (x - meanx_) / stdx_
90      return x
91
92
93  def inverse_transform(x):
94      x = x * stdx_ + meanx_    # non-standardized
95      x = np.exp(x) + xmin - 100.    # inverse log positive
96      return x
97
98
99  xmin = np.array(resampled.min())[0]
100 meanx_ = np.array(resampled.mean())[0]
101 stdx_ = np.array(resampled.std())[0]
102
103 resampled = resampled.apply(transform_data)
104 print("\nDone data transformation")
105
106
107
108 # After transformation and standardization
109 plotGraph(df_=resampled,    # Time series plot
110           label=["Year", "$log($" + "Streamflow" + "$)$"],
111           figsize=(10, 4),
112           indicator=0,
113           title="ses-ts_resampled_after")
114 plotGraph(df_=resampled,    # Histogram
115           label=["Streamflow", ""],
116           figsize=(7, 6),
117           indicator=1,
118           title="ses-histogram_after")
119 plotGraph(df_=resampled,    # Boxplot
120           label=["Year", "Streamflow"],
121           figsize=(14, 4),
122           indicator=2,
123           title="ses-boxplot_after")
124
125
126 """
127 Stationarity check
128 """
129 def stationarity_adf_test(x, alpha=0.05):
130     adftest_res = ADF(x, autolag="AIC")
131     dfout = pd.Series(
132         adftest_res[0:4],
133         index=["ADF statistic", "ADF p-value",
134                 "ADF lags used", "ADF number of obs used"])
```

```
169     for key, value in adftest_res[4].items():
170         dfout["   Critical Value (%s)" % key] = value
171     print(dfout)
172     if dfout["ADF p-value"] > alpha:
173         print("   Result: Non-stationary time series", "\n")
174     else:
175         print("   Result: Stationary time series", "\n")
176
177
178 print("\nChecking stationarity:")
179 stationarity_adf_test(resampled)
180
181
182 """
183 Autocorrelation and Partial autocorrelation
184 """
185 plotGraph(df_=resampled,   # Autocorrelation plot
186           label=["", ""],
187           figsize=(7, 6),
188           indicator=3,
189           title="ses-acorr")
190
191
192 """
193 Data partition and Simple Exponential fitting
194 """
195 trainmask = (resampled.index >= "1981-01-01") & \
196             (resampled.index <= "2002-12-31")
197 testmask = (resampled.index > "2002-12-31")
198 training_set = list(resampled["Streamflow"].loc[trainmask])
199 testing_set = list(resampled["Streamflow"].loc[testmask])
200 print("Created training and testing datasets\n")
201
202 """
203 Single-step predictions
204 alpha = 0.5 (smoothing_parameter)
205 """
206 alpha = 0.5
207 forecasts = list()
208 for i in range(len(testing_set)):
209     modelSES = SimpleExpSmoothing(training_set).fit(
210         smoothing_level=alpha)
211     pred = np.squeeze(modelSES.forecast())
212     forecasts.append(pred)
```

```
207    training_set.append(testing_set[i])
208    print("Obs={:0.04f}, Pred={:0.04f}    {}/{}".format(
209        np.squeeze(inverse_transform(testing_set[i])),
210        np.squeeze(inverse_transform(pred)),
211        i + 1,
212        len(testing_set))
213    )
214 print("\nRoot mean square error:")
215 print("    Testset RMSE={:0.04}".
216        format(MSE(testing_set, forecasts)))
217
218 """
219 Result plots
220 """
221 result = {"Year": resampled["Streamflow"].loc[testmask].index,
222        "Test series":
223            inverse_transform(np.array(testing_set)),
224        "Predicted series":
225            inverse_transform(np.array(forecasts))}
226 df_res = pd.DataFrame.from_dict(data=result)
227 df_res.set_index("Year", inplace=True)
228 plotGraph(df_=df_res,   # Prediction results
229        label=["Year", "Streamflow"],
230        figsize=(10, 4),
231        indicator=4,
232        title="ses-result")
233 print("Done!")
```

The above program implements a time series model using the Simple Exponential Smoothing (SES) model, using Python and several libraries such as pandas, NumPy, matplotlib, statsmodels, sklearn, and Seaborn.

It starts by importing necessary libraries for data handling, plotting, statistical modeling, and performance evaluation. A helper function "plotGraph()" is then defined to generate different types of plots, such as time series (indicator 0), histogram (indicator 1), boxplot (indicator 2), autocorrelation function (ACF), partial autocorrelation function (PACF) (indicator 3), and model predictions (indicator 4).

Next, the script loads a dataset (in this case, the Godavari River flow data) and preprocesses it by checking the format of the date column and resampling the data to downsample the time series data to a 2-monthly mean. It then plots the resampled data using the "plotGraph()" function.

Like any other time series model, our model assumes that the data honors the stationarity and homoscedasticity condition. We next apply a logarithmic transformation and standardization to the data.

After transformation, the program utilizes the Augmented Dickey-Fuller (ADF) test for checking the data's stationarity property. We also plot the autocorrelation and the partial autocorrelation functions of the time series data.

Throughout the execution of the program, we print important information on the console, like, results for stationarity check and RMSE values enabling the progress of steps and outputs easier to understand. The graph, displaying the predicted values and the original time series data, is then output, allowing to evaluate the model's performance.

Chapter 8
Common Hypothesis Testing

Hypothesis testing, also known as significance testing, is widely used in hydrology to determine if the data supports a particular hypothesis. It enables one to make statistical inferences about hydrological phenomena and draw conclusions from data. By designing alternative and null hypotheses, one can test the credibility of model predictions against the observations.

A standard use of hypothesis testing in hydrology is determining whether a difference in the samples from two groups represents the real difference between their populations. For example, scientists might employ t-tests to compare the mean precipitation levels between different seasons or geographical regions. Likewise, Analysis of Variance (ANOVA) tests can be utilized for comparing temperature, river flow rates or precipitation across multiple groups or periods.

Hypothesis testing also comes in handy in assessing the relationships between variables. For instance, one can evaluate the strength of relationships or the degree of correlation between variables, say evapotranspiration and temperature, through Pearson's correlation tests. Using hypothesis testing, one can also confirm if the estimated regression coefficients between variables, say river discharge, rainfall, and temperature, bear any statistical significance.

Moreover, hypothesis testing can also be used for validating the soundness of hydrological models. For example, scientists can employ chi-square goodness-of-fit tests to verify the alignment of observed data frequencies with those predicted by a specific hydrological model. Similarly, a test like Kolmogorov-Smirnov can be used to compare the empirical distribution of model residuals and a theoretical distribution (commonly Gaussian) to validate assumptions and assess model efficacy.

Nevertheless, it is important to exercise caution in applying hypothesis testing. Although it provides invaluable statistical insights, the outcomes intrinsically depend on the sample size and the level of significance chosen. Generally, larger samples enable the detection of smaller statistical effects. Conversely, the level of significance that's commonly opted for (often at 0.05) is somewhat arbitrary and acts as a standard threshold for determining "statistical significance". Moreover, failure to negate the

© The Author(s), under exclusive license to Springer Nature Singapore Pte Ltd. 2024   109
A. Kumar and M. Saharia, *Python for Water and Environment*, Innovations in Sustainable
Technologies and Computing, https://doi.org/10.1007/978-981-99-9408-3_8

null hypothesis should not be misconstrued as a confirmation of it, and statistical significance is not necessarily indicative of practical significance.

So, while hypothesis testing remains a popular tool in hydrological data studies, it needs careful use and understanding.

8.1   One-Way Analysis of Variance

One-way Analysis of Variance (ANOVA) is a statistical technique for checking the difference in the average of three or more groups. It helps us know whether the difference among the groups is significant or not. This test lets us look at hydrologic parameters across different locations, periods, or experimental conditions.

In the context of hydrology, one-way ANOVA might be used to find variations in rainfall, river discharge, groundwater levels, or water quality indicators. The data might originate from separate places, seasons, or test conditions.

To illustrate, consider an investigation exploring the effects of land use on streamflow characteristics across multiple watersheds. We know that watersheds can be placed into various land use categories such as urban, agricultural, and forest. Mean streamflow values can be analyzed and compared among these land use categories using one-way ANOVA. The null hypothesis of ANOVA might suggest no substantial variation in mean streamflow across the different types. Conversely, the alternative hypothesis might suggest a distinct mean streamflow within each land use type. The F-statistic derived from ANOVA measures the ratio between-group and within-group variability. If the F-value is sufficiently large and exceeds the critical value, it indicates that the differences in streamflow between the land use types are significant.

One-way ANOVA is also valuable in hydrological research when analyzing long-term trends or comparing hydrological variables across periods. For instance, it can be used to evaluate the mean annual precipitation or mean river discharge for different decades or climate regimes.

Overall, one-way ANOVA is a useful statistical technique in hydrology for examining differences in hydrological variables among different groups, locations, time periods, or experimental conditions. It provides insights into the variability and significance of these differences, contributing to a better understanding of hydrological processes and informing decision-making in water resources management.

```
1    """
2    Program to demonstrate 1-way ANOVA
3    on the hydrological data.
4    """
5
6    import numpy as np
7    import pandas as pd
8    import statsmodels.api as sm
9    from statsmodels.formula.api import ols
10
11   # Avoiding Pandas warning for column assignment
12   pd.options.mode.chained_assignment = None
13
14   np.random.seed(11)
15
16   """
17   Loading the dataset
18   """
19   # Loading first dataset
20   data1 = pd.read_csv(
21       filepath_or_buffer="../data/Godavari.csv",
22       sep=",",
23       header=0
24   ).dropna()
25   df1 = data1[["Level", "Streamflow"]]  # Retrieving columns
26   df1.loc[:, "River"] = "Godavari"  # Creating new column
27   df1 = df1.iloc[:1000, :]
28
29   # Loading second dataset
30   data2 = pd.read_csv(
31       filepath_or_buffer="../data/Cauvery.csv",
32       sep=",",
33       header=0
34   ).dropna()
35   df2 = data2[["Level", "Streamflow"]]  # Retrieving columns
36   df2.loc[:, "River"] = "Cauvery"  # Creating new column
37   df2 = df2.iloc[:1000, :]
38
39
40   ############## First-part ##############
41   print("\n\nFirst part")
```

```python
# Creating a combined dataframe
df = pd.concat(objs=[df1, df2], axis=0)

"""
Fit a linear model to be used
by the ANOVA routine below
"""
linearModel = ols(
    formula='Streamflow ~ C(River)', data=df).fit()

"""
Perform 1-way ANOVA on the
fitted model
"""
anova_result = sm.stats.anova_lm(linearModel, typ=1)

# Displaying the results
print("\n1. ANOVA result:")
print(anova_result)

# Access the p-values from the ANOVA table then
# compare the p-values to the desired
# significance level (e.g., 0.05)
# finally, print the significant results
alpha = 0.05
p_values = anova_result['PR(>F)'].dropna()
significant_results = p_values > alpha
if significant_results['C(River)']:
    print("\nRiver:")
    print("     Alternate Hypothesis => "
          "No significant difference => "
          "Means are equal")
```

```python
74  else:
75      print("\nRiver:")
76      print("    Null Hypothesis => "
77            "Difference is significant => "
78            "Means are different")
79
80  print("------------------------------------"
81        "------------------------------------")
82
83
84
85
86
87  ############# Second-part #############
88  print("\n\n\n\nSecond part")
89  # Adding noise to the data
90  amp = df1["Streamflow"] * 0.05
91  amp_noisy = amp * np.random.normal(
92      loc=10, scale=2, size=df2.shape[0])
93  amp_noisy = amp_noisy + df2["Streamflow"]
94  df2["Streamflow"] = amp_noisy
95
96  # Creating a combined dataframe
97  dfnoisy = pd.concat(objs=[df1, df2], axis=0)
98
99  """
100 Fit a linear model to be used
101 by the ANOVA routine below
102 """
103 linearModel2 = ols(
104     formula='Streamflow ~ C(River)', data=dfnoisy).fit()
105
106 """
107 Perform 1-way ANOVA on the
108 fitted model
109 """
110 anova_result2 = sm.stats.anova_lm(linearModel2, typ=1)
```

```
111  alpha2 = 0.05
112
113  # Displaying the results
114  print("\n2. ANOVA result:")
115  print(anova_result2)
116  p_values2 = anova_result2['PR(>F)'].dropna()
117  significant_results2 = p_values2 > alpha2
118  if significant_results2['C(River)']:
119      print("\nRiver:")
120      print("        Alternate Hypothesis => "
121            "No significant difference => "
122            "Means are equal")
123  else:
124      print("\nRiver:")
125      print("        Null Hypothesis => "
126            "Difference is significant => "
127            "Means are different")
```

```
●  ●  ●           Terminal

>>> First part
>>> 1. ANOVA result:
>>>                  df          sum_sq          mean_sq              F
↵  PR(>F)
>>> C(River)      1.0   6.500670e+06   6.500670e+06   29.126218
↵  7.586996e-08
>>> Residual   1998.0   4.459329e+08   2.231896e+05            NaN
↵  NaN
>>> River:
>>>        Null Hypothesis => Difference is significant => Means are
↵  different
>>>
↵  --------------------------------------------------------------------
>>> Second part
>>> 2. ANOVA result:
>>>                  df          sum_sq          mean_sq              F
↵  PR(>F)
>>> C(River)      1.0   4.085371e+04   40853.708320   0.133885
↵  0.714476
>>> Residual   1998.0   6.096690e+08   305139.619801          NaN
↵  NaN
```

This program demonstrates the application of one-way ANOVA (analysis of variance) on hydrological data from two rivers, Godavari and Cauvery. The primary goal of this analysis is to test if there is a significant difference in streamflow between the two rivers.

First, the program imports necessary libraries such as NumPy, Pandas, and Statsmodels. Then, it loads two datasets corresponding to the Godavari and

Cauvery Rivers, each dataset having variables "Level" and "Streamflow". The datasets are loaded using Pandas, and any missing values are dropped. A new column, "River", is added to the dataframes to label data of the respective river.

The program then combines these two datasets into one dataframe to facilitate the comparison of streamflow between the two rivers. Following this, a linear model is fit to the combined data, with the streamflow as the dependent variable and the river as the independent categorical variable. The "ols" function from the statsmodels.formula.api module fits this linear model.

After the model is fitted, a one-way ANOVA test is performed. The ANOVA test results, including the F-statistic and p-value, are printed to the console. A p-value less than 0.05 is typically considered evidence of a significant difference.

The program also performs a separate analysis where some noise is added to the streamflow data of the Cauvery River. This noise represents some random variation in the data. The noisy data are then used to fit a new linear model and perform a one-way ANOVA test in the same manner as before. The results of this second analysis allow one to examine how the added noise affects the outcome of the ANOVA test.

In both cases, the p-values from the ANOVA table are accessed and compared to a significance level of 0.05 to determine if the means of the streamflow from the two rivers are significantly different. If the p-value is less than 0.05, it concludes that the means are different; otherwise, it suggests that there is no significant difference in the means.

8.2 Two-Way Analysis of Variance

Two-way ANOVA (analysis of variance) is a statistical test used to examine the effects of two independent categorical variables (factors) on a continuous dependent variable. ANOVA helps in the assessment of the main effects of each factor as well as the interaction between the factors. In hydrology, two-way ANOVA can be utilized in various scenarios where multiple factors influence a hydrological response.

In the context of hydrology, a common application of two-way ANOVA is the analysis of the impact of climate (e.g., precipitation levels) and land use (e.g., agricultural practices) on hydrological variables such as streamflow or water quality parameters. The effect of climate can vary with different seasons or zones experiencing different precipitation levels. On the other hand, land use may be categorized into multiple types based on cover type and management. With two-way ANOVA, we can analyze the independent effect of the two factors on hydrological parameters. A possible extension of the ANOVA model would be to study the mixed effect.

In two-way ANOVA, the total sum of squares of the residuals can be attributed to three contributing factors: 1. effect due to factor A (SSA), 2. effect due to factor B (SSB), and 3. the combined effect due to A and B (SSAB). The ANOVA results can be interpreted by analyzing the interaction, degrees of freedom, and the mean squares error. An accompanying F-statistic is also computed to evaluate the significance of the factors, both individual and combined.

Thus, by analyzing hydrological data using two-way ANOVA, one can simultaneously gauge the independent and combined effects of multiple variables. It provides a way to ascertain if the response is due to individual or combined effects. This type of analysis is invaluable in comprehending the complex dynamics inherent in hydrologic processes, consequently helping in decision-making and framing adaptation strategies to reduce hydrologic risks.

```python
"""
Program to demonstrate 2-way ANOVA
on the hydrological data.
"""

import numpy as np
import pandas as pd
import statsmodels.api as sm
from statsmodels.formula.api import ols

# Avoiding Pandas warning for column assignment
pd.options.mode.chained_assignment = None

np.random.seed(11)

"""
Loading the dataset
"""
# Loading first dataset
data1 = pd.read_csv(
    filepath_or_buffer="../data/Godavari.csv",
    sep=",",
    header=0
).dropna()
df1 = data1[["Level", "Streamflow"]]  # Retrieving columns
df1.loc[:, "River"] = "Godavari"  # Creating new column
df1 = df1.iloc[:1000, :]

# Loading second dataset
data2 = pd.read_csv(
    filepath_or_buffer="../data/Cauvery.csv",
    sep=",",
    header=0
).dropna()
df2 = data2[["Level", "Streamflow"]]  # Retrieving columns
df2.loc[:, "River"] = "Cauvery"  # Creating new column
df2 = df2.iloc[:1000, :]
```

```
37   ############## First-part ##############
38   print("\n\nFirst part")
39   # Creating a combined dataframe
40   df = pd.concat(
41       objs=[df1, df2],
42       axis=0
43   )
44
45
46   """
47   Fit a linear model to be used
48   by the ANOVA routine below
49   """
50   linearModel = ols(
51       formula='Streamflow ~ Level + C(River)',
52       data=df).fit()
53
54
55   """
56   Perform 2-way ANOVA on the
57   fitted model
58   """
59   anova_result = sm.stats.anova_lm(
60       linearModel,
61       typ=2
62   )
63   alpha = 0.05
64
65   # Displaying the results
66   print("\n1. ANOVA result:")
67   print(anova_result)
68
69   # Access the p-values from the ANOVA table then
70   # compare the p-values to the desired
71   # significance level (e.g., 0.05)
72   # finally, print the significant results
73   p_values = anova_result['PR(>F)'].dropna()
74   significant_results = p_values > alpha
```

```python
78  if significant_results['C(River)']:
79      print("\nRiver:")
80      print("     Alternate Hypothesis => "
81          "No significant difference => "
82          "Means are equal")
83  else:
84      print("\nRiver:")
85      print("     Null Hypothesis => "
86          "Difference is significant => "
87          "Means are different")
88  if significant_results['Level']:
89      print("\nLevel:")
90      print("     Alternate Hypothesis => "
91          "No significant difference => "
92          "Means are equal")
93  else:
94      print("\nLevel:")
95      print("     Null Hypothesis => "
96          "Difference is significant => "
97          "Means are different")
98
99  print("------------------------------------"
100         "------------------------------------")
101
102
103  ############# Second-part #############
104  print("\n\n\n\nSecond part")
105  # Adding noise to the data
106  amp = df2["Level"] * 0.05
107  amp_noisy = amp * np.random.normal(
108      loc=10, scale=2, size=df2.shape[0])
109  amp_noisy = amp_noisy + df1["Level"]
110  df2["Level"] = amp_noisy
111
112  # Creating a combined dataframe
113  dfn = pd.concat(
114      objs=[df1, df2],
115      axis=0
116  )
117
118
119  """
120  Fit a linear model to be used
121  by the ANOVA routine below
122  """
```

```python
linearModel2 = ols(
    formula='Streamflow ~ Level + C(River)',
    data=dfn).fit()



"""
Perform 2-way ANOVA on the
fitted model
"""
anova_result2 = sm.stats.anova_lm(
    linearModel2,
    typ=2
)
alpha2 = 0.05

# Displaying the results
print("\n2. ANOVA result:")
print(anova_result2)
p_values2 = anova_result2['PR(>F)'].dropna()
significant_results2 = p_values2 > alpha2

if significant_results2['C(River)']:
    print("\nRiver:")
    print("      Alternate Hypothesis => "
          "No significant difference => "
          "Means are equal")
else:
    print("\nRiver:")
    print("      Null Hypothesis => "
          "Difference is significant => "
          "Means are different")

if significant_results2['Level']:
    print("\nLevel:")
    print("      Alternate Hypothesis => "
          "No significant difference => "
          "Means are equal")
else:
    print("\nLevel:")
    print("      Null Hypothesis => "
          "Difference is significant => "
          "Means are different")
```

```
>>> First part
>>> 1. ANOVA result:
>>>                 sum_sq         df              F   PR(>F)
>>> C(River)   3.633828e+08       1.0    8740.364782     0.0
>>> Level      3.629071e+08       1.0    8728.921893     0.0
>>> Residual   8.302577e+07    1997.0            NaN     NaN
>>> River:
>>>      Null Hypothesis => Difference is significant => Means are
↵   different
>>> Level:
>>>      Null Hypothesis => Difference is significant => Means are
↵   different
>>>
↵   -------------------------------------------------------------------
>>> Second part
>>> 2. ANOVA result:
>>>                 sum_sq         df              F    PR(>F)
>>> C(River)   1.907740e+06       1.0    8.552940    0.003489
>>> Level      5.003516e+05       1.0    2.243218    0.134359
>>> Residual   4.454325e+08    1997.0         NaN         NaN
>>> River:
>>>      Null Hypothesis => Difference is significant => Means are
↵   different
>>> Level:
>>>      Alternate Hypothesis => No significant difference => Means
↵   are equal
```

The Python program provided above demonstrates a two-way Analysis of Variance (ANOVA) on hydrological data from two rivers—Godavari and Cauvery.

The first part of the program loads two separate datasets related to the two rivers. The Pandas library is used to read the data from CSV files, drop any rows with missing data, and create new dataframes with selected columns. Each dataframe is limited to the first 1000 rows of data and a new column, "River", is added to identify the source of the data. These two dataframes are then combined into a single dataframe.

In the next step, the program fits a linear model to the data using the ordinary least squares (ols) method from the statsmodels library, taking "Streamflow" as the dependent variable and "Level" and "River" as independent variables.

After fitting the linear model, a two-way ANOVA is performed on the model. The ANOVA results are displayed, showing the sum of squares, degrees of freedom, mean square, F-statistic, and p-value for each factor and the interaction between them.

Next, the program evaluates the p-values from the ANOVA table to determine statistical significance. If the p-value is less than the significance level (alpha = 0.05), the null hypothesis is rejected, suggesting a significant difference in streamflow between the rivers or at different levels.

The second part of the program introduces some noise into the "Level" data for the Cauvery River by multiplying the "Level" data with random variates from a normal distribution and adding it to the "Level" data from the Godavari River. The new noisy data is combined with the Godavari data into a new dataframe.

Similar to the first part of the program, a linear model is fitted to the noisy data and a two-way ANOVA is performed. The results are again displayed, and the p-values are evaluated to determine if there is a significant difference in the means. The results of this analysis will show whether the introduced noise affects the conclusions of the ANOVA.

8.3 t-Test

A t-test is a statistical hypothesis test that determines whether there is a significant difference between the means of two groups. It assumes that the populations follow a normal distribution and have the same variance. It is widely used in research to compare two independent or related samples.

```python
"""
Program demonstrating t-Test
using scipy
"""

import numpy as np
import pandas as pd
from scipy import stats

# Avoiding Pandas warning for column assignment
pd.options.mode.chained_assignment = None

np.random.seed(11)

"""
Loading the dataset
"""
# Loading first dataset
data1 = pd.read_csv(
    filepath_or_buffer="../data/Godavari.csv",
    sep=",",
    header=0
).dropna()
df1 = data1[["Level", "Streamflow"]]   # Retrieving columns
df1.loc[:, "River"] = "Godavari"   # Creating new column
df1 = df1.iloc[:1000, :]

# Loading second dataset
data2 = pd.read_csv(
    filepath_or_buffer="../data/Cauvery.csv",
    sep=",",
    header=0
).dropna()
df2 = data2[["Level", "Streamflow"]]   # Retrieving columns
df2.loc[:, "River"] = "Cauvery"   # Creating new column
```

```python
36   df2 = df2.iloc[:1000, :]
37
38   # Performing the t-Test (First part)
39   alpha = 0.05
40   t_statistics = stats.ttest_ind(
41       a=df1["Streamflow"], b=df2["Streamflow"],
42       equal_var=False
43   )
44   print(t_statistics)
45   print("\n1. Different samples:")
46   print("      p={:0.010f}".format(t_statistics[1]))
47   if t_statistics[1] < alpha:
48       print("-> The sample means are different")
49   else:
50       print("-> The sample means are same")
51
52
53   print("\n------------------------------------"
54        "-------------------------------------\n")
55
56   # Modifying data to be similar to the first (Second part)
57   amp = df1["Streamflow"] * 0.05    # 5% of the amplitude
58   amp_noisy = amp * np.random.normal(
59       loc=10, scale=2, size=df2.shape[0])
60   amp_noisy = amp_noisy + df2["Streamflow"]
61   df2["Streamflow"] = amp_noisy
62   t_statistics2 = stats.ttest_ind(
63       a=df1["Streamflow"], b=df2["Streamflow"],
64       equal_var=False
65   )
66   print(t_statistics2)
67   print("\n2. Modified samples:")
68   print("      p={:0.010f}".format(t_statistics2[1]))
69   if t_statistics2[1] < alpha:
70       print("-> The sample means are different")
71   else:
72       print("-> The sample means are same")
```

```
●  ●  ●          Terminal
>>> Ttest_indResult(statistic=5.396871142823051,
↳   pvalue=8.214052581068897e-08)
>>> 1. Different samples:
>>>      p=0.0000000821
>>>        -> The sample means are different
>>>
↳   ------------------------------------------------------------
>>> Ttest_indResult(statistic=0.3659033970931655,
↳   pvalue=0.7144806864198379)
>>> 2. Modified samples:
>>>      p=0.7144806864
>>>        -> The sample means are same
```

This Python program demonstrates a t-test using the SciPy library. It is designed to compare the streamflow of the Godavari and the Cauvery Rivers in India. The main aim of this program is to test whether there is a significant difference in the streamflows of these two rivers.

The program starts by importing the necessary libraries: NumPy, Pandas, and SciPy. A seed is set for NumPy's random number generator to ensure reproducibility of the results.

Next, the program loads two datasets, one for each river. The datasets are CSV files and are loaded using the Pandas function "pd.read_csv()". Columns related to the "Level" and "Streamflow" are retrieved from both datasets. An additional column named "River" is created to store the name of each river. Only the first 1000 rows of each dataset are kept for comparison of the two rivers.

The first part of the t-test is then performed using the "ttest_ind()" function from the stats module of SciPy. This function calculates the test's t-statistic and p-value assuming unequal variances between the two rivers' streamflows. The test result is printed, and if the p-value is less than the chosen significance level (0.05 in this case), it is concluded that the streamflows of the two rivers have significantly different means.

Next, in the second part, the streamflow data of the Cauvery River is modified to mimic the streamflow of the Godavari River. This is done by adding a random noise that amounts to 5% of the Godavari River's streamflow to the Cauvery River's streamflow.

A second t-test is performed on the modified data. The aim here is to see if the modification has resulted in similarities in the streamflows of the two rivers, as reflected by the p-value. If the p-value is greater than the significance level, it suggests that the means of the streamflows are statistically similar, hence indicating that the modification was successful.

8.4 F-Test

The F-test is a statistical test used to compare the variances of two or more groups to see if they are equal. It is often used in the context of ANOVA (analysis of variance), regression analysis, or to compare nested models. A significant F-test suggests the group variances are unequal.

```python
"""
Program to  perform F-Test
using scipy in Python
"""

import numpy as np
import pandas as pd
from scipy import stats

# Avoiding Pandas warning for column assignment
pd.options.mode.chained_assignment = None

np.random.seed(11)

"""
Loading the dataset
"""
# Loading first dataset
data1 = pd.read_csv(
    filepath_or_buffer="../data/Godavari.csv",
    sep=",",
    header=0
).dropna()
df1 = data1[["Level", "Streamflow"]]  # Retrieving columns
df1.loc[:, "River"] = "Godavari"  # Creating new column
df1 = df1.iloc[:1000, :]

# Loading second dataset
data2 = pd.read_csv(
    filepath_or_buffer="../data/Cauvery.csv",
    sep=",",
    header=0
).dropna()
df2 = data2[["Level", "Streamflow"]]  # Retrieving columns
df2.loc[:, "River"] = "Cauvery"  # Creating new column
df2 = df2.iloc[:1000, :]
```

```
36   """
37   Define a function for F-Test
38   """
39   def F_Test(a, b):
40       # Computing variances
41       var1 = np.var(a, ddof=1)
42       var2 = np.var(b, ddof=1)
43       fstat = np.divide(var1, var2)
44
45       # Getting DOFs of samples
46       dof_numerator, dof_denominator = \
47           a.shape[0]-1, b.shape[0]-1
48
49       # Computing p-value using f distribution's cdf
50       p_val = 1. - stats.f.cdf(
51           x=fstat, dfn=dof_numerator, dfd=dof_denominator
52       )
53
54       # Returning results
55       return fstat, p_val
56
57
58   # Performing the F-Test (First part)
59   alpha = 0.05
60   f_statistics = F_Test(
61       a=df1["Streamflow"], b=df2["Streamflow"]
62   )
63   print(f_statistics)
64   print("\n1. Different samples:")
65   print("      p={:0.010f}".format(f_statistics[1]))
66   if f_statistics[1] < alpha:
67       print("      -> The sample variances are different")
68   else:
69       print("      -> The sample variances are same")
70
71
72   print("\n------------------------------------"
73         "------------------------------------\n")
74
75   # Modifying data to be similar to the first (Second part)
```

```python
79   amp = df1["Streamflow"] * 0.08   # 8% of the amplitude
80   amp_noisy = amp * np.random.normal(
81       loc=10, scale=2, size=df2.shape[0])
82   amp_noisy = amp_noisy + df2["Streamflow"]
83   df2["Streamflow"] = amp_noisy
84   f_statistics2 = F_Test(
85       a=df1["Streamflow"], b=df2["Streamflow"]
86   )
87   print(f_statistics2)
88   print("\n2. Modified samples:")
89   print("     p={:0.010f}".format(f_statistics2[1]))
90   if f_statistics2[1] < alpha:
91       print("-> The sample variances are different")
92   else:
93       print("-> The sample variances are same")
```

```
  ●  ●  ●        Terminal

>>> (12.149521092840168, 1.1102230246251565e-16)
>>> 1. Different samples:
>>>      p=0.0000000000
>>>        -> The sample variances are different
>>>

⮑ --------------------------------------------------------------
>>> (1.033911521888177, 0.29913519223758955)
>>> 2. Modified samples:
>>>      p=0.2991351922
>>>        -> The sample variances are same
```

The given Python program performs an F-test on two sets of data—pertaining to two different rivers, the Godavari and Cauvery. This is a statistical test that compares the variances of the two samples to ascertain if they are significantly different.

The program first imports the necessary libraries: NumPy for numerical computations, Pandas for data manipulation, and SciPy for statistical functions. To avoid a specific type of warning generated by Pandas, a setting is adjusted. A seed for the random number generator in NumPy is set to ensure that the results are reproducible.

The program then loads two datasets from .csv files that contain data about the rivers Godavari and Cauvery, respectively. Any missing data points are dropped to ensure the accuracy of the analysis. From each dataset, the "Level" and "Streamflow" columns are extracted, and a new column "River" is added to label the data. The first 1000 rows of each dataset are retained for further processing.

Next, the program defines a function for performing the F-test. This function computes the variances of the two input datasets and calculates the F-statistic as the ratio of the variances. The function then uses the cumulative distribution function (CDF) of the F-distribution to calculate the p-value associated with the observed F-statistic.

The F-test is then performed on the "Streamflow" data of the two rivers, and the results are printed. The test provides a p-value which, if smaller than a chosen significance level (alpha), indicates that the variances of the two datasets are significantly different. The significance level is set at 0.05.

In the second part of the program, the "Streamflow" data for the Cauvery River is modified by adding noise. This noise is generated based on 8% of the amplitude of the "Streamflow" data of the Godavari River. An F-test is again performed on the "Streamflow" data of the two rivers but with the modified data for the Cauvery River. The results of this test are then printed in the same manner as before.

The second test aims to demonstrate how the F-test can detect changes in the variance, even when the change is introduced artificially. By comparing the results of the two tests, one can see how the F-test responds to changes in the data.

8.5 The Kolmogorov-Smirnov Test

The Kolmogorov-Smirnov test is a non-parametric statistical test used to compare a sample distribution with a theoretical distribution (one-sample K-S test) or to compare two-sample distributions (two-sample K-S test). It calculates the maximum distance between the empirical cumulative distribution functions (ECDF) of the two samples or distributions.

```
1    """
2    Program to do a Kolmogorov-Smirnov
3    Test using scipy in Python
4    """
5
6    import numpy as np
7    import pandas as pd
8    from scipy import stats
9
10   # Avoiding Pandas warning for column assignment
11   pd.options.mode.chained_assignment = None
12
13   np.random.seed(11)
14
15   """
16   Loading the dataset
17   """
18   # Loading first dataset
19   data1 = pd.read_csv(
20       filepath_or_buffer="../data/Godavari.csv",
21       sep=",",
22       header=0
23   ).dropna()
24   # Loading second dataset
25   data2 = pd.read_csv(
26       filepath_or_buffer="../data/Cauvery.csv",
27       sep=",",
28       header=0
29   ).dropna()
30
31   df1 = data1[["Level", "Streamflow"]]   # Retrieving columns
32   df1.loc[:, "River"] = "Godavari"   # Creating new column
33   df1 = df1.iloc[:1000, :]
34
35   df2 = data2[["Level", "Streamflow"]]   # Retrieving columns
36   df2.loc[:, "River"] = "Cauvery"   # Creating new column
37   df2 = df2.iloc[:1000, :]
```

```
37   """
38   Kolmogorov-Smirnov Test
39   """
40   alpha = 0.05  # significance
41   # First case - different datasets
42   KS_result = stats.ks_2samp(
43       data1=df1["Streamflow"],
44       data2=df2["Streamflow"]
45   )
46   print("\n1. Kolmogorov-Smirnov Test")
47   print("   p-value={}".format(KS_result[1]))
48   if KS_result[1] < alpha:
49       print("   The two distributions are different")
50   else:
51       print("   The two distributions are same")
52
53   print("\n---------------------------------"
54         "-------------------------------------\n")
55
56   # Second case - Similar datasets
57   chr = 0.8  # character
58   amp_noisy = chr*df1["Streamflow"] + \
59               (1.-chr)*df2["Streamflow"]*np.random.normal(
60       loc=0.08, scale=0.01, size=df2.shape[0])
61   df2["Streamflow"] = amp_noisy
62   KS_result2 = stats.ks_2samp(
63       data1=df1["Streamflow"],
64       data2=df2["Streamflow"]
65   )
66   print("\n2. Kolmogorov-Smirnov Test")
67   print("   p-value={}".format(KS_result2[1]))
68   if KS_result2[1] < alpha:
69       print("The two distributions are different")
70   else:
71       print("The two distributions are same")
```

```
 ● ● ●          Terminal
>>> 1. Kolmogorov-Smirnov Test
>>>    p-value=5.261321340466556e-51
>>>    The two distributions are different
>>>
╰  ------------------------------------------------------------------
>>> 2. Kolmogorov-Smirnov Test
>>>    p-value=0.06153429181321559
>>>    The two distributions are same
```

The given Python script carries out a Kolmogorov-Smirnov (K-S) test using the SciPy library. This test compares the distributions of two datasets to ascertain whether they are significantly different or not.

Firstly, the script imports necessary libraries and modules like NumPy, Pandas, and Stats from SciPy. The Pandas warning for column assignment is avoided by setting a particular Pandas option. The NumPy random seed is also set to ensure the consistency of the pseudo-random numbers generated in the program.

Following this, two separate datasets named "Godavari.csv" and "Cauvery.csv" are loaded. They presumably contain river data about water levels and streamflow. Any missing values in the datasets are discarded using the "dropna()" function. Two specific columns, "Level" and "Streamflow", from the datasets are retrieved and stored in new dataframes. The dataframes are also appended with a new column named "River", which labels the data by the respective river name.

A Kolmogorov-Smirnov test is then performed on the "Streamflow" columns of both dataframes. The significance level is set at 0.05. If the p-value returned by the test is less than this significance level, it implies that the two distributions are significantly different. Otherwise, they are considered to be the same. The results are then printed to the console.

Subsequently, the script simulates a case where the two distributions are likely to be more similar. It generates a noisy version of the "Streamflow" column of the first dataframe by adding Gaussian noise to 80% of the "Streamflow" data of the second dataframe. This new "noisy" dataset replaces the original "Streamflow" column of the second dataframe.

Finally, the Kolmogorov-Smirnov test is repeated on the "Streamflow" columns of the first dataframe and the newly modified second dataframe. Once again, if the p-value is less than the significance level, it implies that the two distributions are significantly different, and the script prints this result. Otherwise, it reports that the two distributions are the same.

8.6 Mann-Whitney Test

The Mann-Whitney test, also known as the Wilcoxon rank-sum test, is another non-parametric statistical test used to determine whether two independent samples were drawn from a population with the same distribution. It is particularly useful when the data does not meet the assumptions required for a t-test.

```python
"""
Program to do a Mann-Whitney
Test using scipy in Python
"""

import numpy as np
import pandas as pd
from scipy import stats

# Avoiding Pandas warning for column assignment
pd.options.mode.chained_assignment = None

np.random.seed(11)

"""
Loading the dataset
"""
# Loading first dataset
data1 = pd.read_csv(
    filepath_or_buffer="../data/Godavari.csv",
    sep=",",
    header=0
).dropna()
# Loading second dataset
data2 = pd.read_csv(
    filepath_or_buffer="../data/Cauvery.csv",
    sep=",",
    header=0
).dropna()

df1 = data1[["Level", "Streamflow"]]  # Retrieving columns
df1.loc[:, "River"] = "Godavari"  # Creating new column
df1 = df1.iloc[:1000, :]

df2 = data2[["Level", "Streamflow"]]  # Retrieving columns
df2.loc[:, "River"] = "Cauvery"  # Creating new column
df2 = df2.iloc[:1000, :]
```

```
37  """
38  Mann-Whitney Test
39  """
40  alpha = 0.05  # significance
41  MW_statistic, p_val = stats.mannwhitneyu(
42      x=df1["Streamflow"], y=df2["Streamflow"],
43      method="auto"
44  )
45  print("\n1. Mann-Whitney Test")
46  print("   p-value={}".format(p_val))
47  if p_val < alpha:
48      print("   The two distributions are different")
49  else:
50      print("   The two distributions are same")
51
52  print("\n-----------------------------------"
53      "-----------------------------------\n")
54
55  # Second case - Similar datasets
56  chr = 0.8  # character
57  amp_noisy = chr*df1["Streamflow"] + \
58              (1.-chr)*df2["Streamflow"]*np.random.normal(
59      loc=0.08, scale=0.01, size=df2.shape[0])
60  df2["Streamflow"] = amp_noisy
61  MW_statistic2, p_val2 = stats.mannwhitneyu(
62      x=df1["Streamflow"], y=df2["Streamflow"],
63      method="auto"
64  )
65  print("\n2. Mann-Whitney Test")
66  print("   p-value={}".format(p_val2))
67  if p_val2 < alpha:
68      print("The two distributions are different")
69  else:
70      print("The two distributions are same")
```

```
●  ●  ●        Terminal

>>> 1. Mann-Whitney Test
>>>    p-value=8.062468634889823e-16
>>>    The two distributions are different
>>>
    ----------------------------------------------------------------
>>> 2. Mann-Whitney Test
>>>    p-value=0.3962091294974023
>>>    The two distributions are same
```

The adjoining Python program executes the Mann-Whitney test on two datasets. It starts by importing the essential libraries: NumPy, Pandas, and the Stats module from SciPy. The line, "pd.mode.chained_assignment=None", ensures that the warnings corresponding to column assignments in a Pandas dataframe are suppressed. A random seed is also specified to ensure the reproducibility of the results.

Next, we load the two datasets from two external csv files: one for the Godavari River and another for the Cauvery River. The "dropna()" function from within the Pandas library is used to drop any rows containing invalid entries in the dataframes, creating two datasets named "df1" and "df2". Only the "Level" and "Streamflow" columns are extracted for processing. A new column "River", is added to both dataframes to label the river each dataset represents. The program then limits each dataframe to the first 1000 rows.

The Mann-Whitney test is then applied using the "mannwhitneyu()" function from SciPy's stats module. The test is performed on the "Streamflow" data of both rivers. The significance level is set at 0.05. The test returns a statistic and a p-value. The p-value is compared to the significance level to decide whether the two distributions are the same or different. If the p-value is less than the significance level, the program concludes that the distributions are different, otherwise, they are deemed to be the same.

Following this, the program creates new "Streamflow" data for the Cauvery River (df2) as a noisy version of the "Streamflow" data of the Godavari River (df1). This new "Streamflow" data is a mixture of the Godavari River's "Streamflow" and a normal random noise scaled by the Cauvery's "Streamflow". A second Mann-Whitney test is conducted on this new data and the Godavari's "Streamflow" data, again comparing the p-value with the significance level to draw a conclusion about the similarity of the two distributions.

Chapter 9
Uncertainty Estimation

Estimating uncertainty is an important part of hydrological data modeling. They offer a measure of confidence in model predictions and aid in decision-making under risk.

Hydrological models are often complex, aiming to represent the intricate relationships among various environmental and human factors. They rely on input data, such as precipitation, temperature, soil characteristics, and land use, which can have uncertainties due to measurement errors, data gaps, or spatial and temporal variability. But these models incorporate simplifications of complex physical process representations in order to apply them in real-world situations, which introduces further uncertainty.

Estimates of uncertainty provide an understanding of how different elements of uncertainty affect the outcomes generated by the model. We can pinpoint the likely spectrum within which the actual value may reside by quantifying uncertainty rooted in the utilized data and model. For instance, a prediction concerning river flow could be supplemented by an uncertainty estimate, outputting not a singular value but a probable range of outcomes.

Uncertainty quantification also contributes to a better understanding of the dependability of the model's predictions. In other words, it helps to communicate the confidence level we can place in these predictions. It measures the degree of certainty or risk associated with decisions based on model outcomes.

Additionally, understanding the sources and magnitudes of uncertainties can guide model refinement and data collection strategies. For instance, if a significant portion of the uncertainty in a model output arises from uncertain precipitation data, it may be worth investing in more accurate precipitation measurement techniques or refining the representation of precipitation processes in the model.

Uncertainty quantification also plays a crucial role in risk management and decision-making. For instance, in flood risk management, knowing the uncertainty associated with a prediction of river discharge can help in setting appropriate safety margins when designing flood defenses.

© The Author(s), under exclusive license to Springer Nature Singapore Pte Ltd. 2024   135
A. Kumar and M. Saharia, *Python for Water and Environment*, Innovations in Sustainable Technologies and Computing, https://doi.org/10.1007/978-981-99-9408-3_9

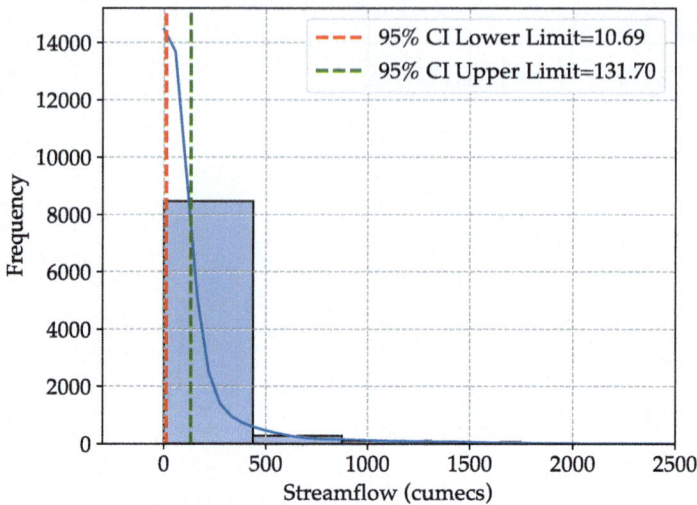

Fig. 9.1 Non-parametric interval estimate

Furthermore, uncertainty estimates are important for scenario analysis, studying the impacts of climate change. Given the uncertainties in future greenhouse gas emissions and climate models, hydrological predictions for future conditions are inherently uncertain. Quantifying these uncertainties can help policymakers understand the range of possible outcomes and develop robust adaptation strategies.

Uncertainty estimates and quantification in hydrology data modeling are of paramount importance. They not only improve our understanding of the confidence we can place in model predictions but also guide model improvement efforts, support risk management, and aid in decision-making under uncertain conditions. While it is impossible to completely eliminate uncertainty, recognizing and quantifying is a crucial part of hydrological modeling.

9.1   Interval Estimates

9.1.1   Non-parametric Interval Estimate

Non-parametric interval estimation involves estimating the range within which a population parameter, like the median or a percentile, is likely to fall, without assuming that the data follows a specific distribution (Fig. 9.1).

```
1   """
2   Program for non-parametric interval
3   estimate on 1D stream flow data
4   """
5   import pandas as pd
6   from scipy.stats import (norm, binom, tmean, tstd,
7                            t, nct, bootstrap, mstats)
8   import numpy as np
9
10  """
11  Load dataset
12  """
13  data = pd.read_csv(
14      filepath_or_buffer="../data/Godavari.csv",
15      sep=",",
16      header=0).dropna()
17  print("\nChecking data:")
18  try:
19      data["time"] = pd.to_datetime(
20          data['time'], infer_datetime_format=True)
21      print("   Date format is okay!\n")
22  except ValueError:
23      print("   Encountered error!\n")
24      pass
25  x = data[["Streamflow"]].to_numpy().squeeze()
26  del data
27  print("Read data file")
28
29  """
30  Non-parametric interval estimate
31  """
32  print("\n\nDoing non-parametric interval estimate")
33  qnt = binom.ppf(q=[0.025, 0.975],
34                  n=len(x), p=0.5,
35                  loc=0)
36  qnt = np.cast[int](qnt - 1)
37  print("Quantile values = ({}, {})".format(x[qnt[0]],
38                                             x[qnt[1]]))
39
40  print("\nDone!")
```

```
●  ●  ●        Terminal

>>> Checking data:
>>>     Date format is okay▌
>>> Read data file
>>> Doing non-parametric interval estimate
>>> Quantile values = (10.69, 131.7)
>>> Done▌
```

The adjoining program performs a non-parametric interval estimate with 1D stream flow data. The code starts by importing the essential Python libraries like: NumPy for matrix operations, scipy.stats for statistical computations, and Pandas from data manipulation.

The data is loaded from an external .csv file named as "Godavari.csv". For this, Pandas' "read_csv()" function is used. The "dropna()" function eliminates any missing data or null values.

After loading the data, we check for the format of the "time" column. We use the "to_datetime()" function from the Pandas library for this. Upon successful conversion, the program prints a message on the console; otherwise, it throws a value error. In either of the cases, the execution continues.

The necessary data is extracted from the "Streamflow" column, and the dataframe is deleted upon extraction. Subsequently, the dataframe is converted into a 1D NumPy array with the "squeeze()" function, aiding in the statistical calculation in the following steps.

Next, the non-parametric interval estimate is then performed with the help of the binomial percentile point function (ppf). The confidence interval and the success probability for this are specified as 95% (0.025 and 0.975) and 0.5. As a result, we get the two quantile values representing the bounds corresponding to the confidence interval.

Finally, the program prints these quantile values to the console and a message indicating the completion of the interval estimate.

9.2   Confidence Intervals

9.2.1   For Median

9.2.1.1   Bootstrap Confidence Interval Estimate

The bootstrap confidence interval estimate for the median is a non-parametric method that generates many resamples of the observed data, each with replacement. The median of each resample is computed, resulting in a distribution of medians. The confidence interval is then derived from this distribution, capturing the central tendency of the data (Fig. 9.2).

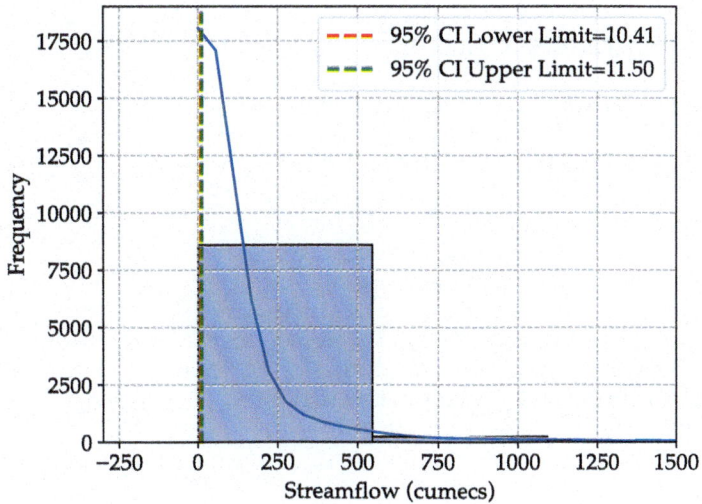

Fig. 9.2 Bootstrap confidence interval estimate

```
1   """
2   Program for Bootstrap confidence
3   interval estimate on 1D stream flow data
4   """
5   import pandas as pd
6   from scipy.stats import (norm, binom, tmean, tstd,
7                            t, nct, bootstrap, mstats)
8   import numpy as np
9
10  """
11  Load dataset
12  """
13  data = pd.read_csv(
14      filepath_or_buffer="../data/Godavari.csv",
15      sep=",",
16      header=0).dropna()
17  print("\nChecking data:")
18  try:
19      data["time"] = pd.to_datetime(
20          data['time'], infer_datetime_format=True)
21      print("   Date format is okay!\n")
22  except ValueError:
23      print("   Encountered error!\n")
24      pass
25  x = data[["Streamflow"]].to_numpy().squeeze()
26  del data
27  print("Read data file")
```

```python
1   """
2   Bootstrap confidence interval estimate
3   """
4   print("\n\nDoing Bootstrap CI estimate")
5   x = (x,)
6   ci = bootstrap(data=x,
7                   statistic=np.median,
8                   confidence_level=0.95,
9                   n_resamples=10000,
10                  method="percentile")
11  lower_bound = ci.confidence_interval.low
12  upper_bound = ci.confidence_interval.high
13  print(f"Lower Bound: {lower_bound}")
14  print(f"Upper Bound: {upper_bound}")
15
16  print("\nDone!")
```

```
●  ●  ●          Terminal

>>> Checking data:
>>>    Date format is okay!
>>> Read data file
>>> Doing Bootstrap CI estimate
>>> Lower Bound: 10.41
>>> Upper Bound: 11.5
>>> Done!
```

This Python program computes the Bootstrap confidence interval estimate on a one-dimensional streamflow dataset.

Essential libraries are first loaded in the program. The "read_csv()" function of the Pandas library is called to load the data and its dropna() function is used to discard any rows with missing or invalid (NaNs) values.

The "time" column in the loaded dataframe is then converted to a more suitable datetime object printing a message if the conversion is successful else throwing an error. The program continues to execute after the conversion.

The streamflow data are then extracted from the "Streamflow" column into a NumPy array followed by a freeing up of the memory. The program uses the "scipy.stats" library to compute the bootstrap interval by resampling the data multiple times (10000 times in this case) and computing the desired statistic, the median, each time. This way it builds a distribution of the statistic thus helping to compute the 95% confidence interval. The lower and upper bounds of the computed confidence interval are then printed in the console indicating the likely range where the median in expected to fall with a 95% confidence.

9.2.2 For Mean

9.2.2.1 Symmetric Confidence Interval Estimate

A symmetric confidence interval estimate for the mean is a range within which the true population mean is expected to fall, with a certain degree of confidence. It is calculated using the sample mean plus and minus a margin of error, which is determined by the standard error of the mean, and a critical value from a statistical distribution (commonly the t- or z-distribution) (Fig. 9.3).

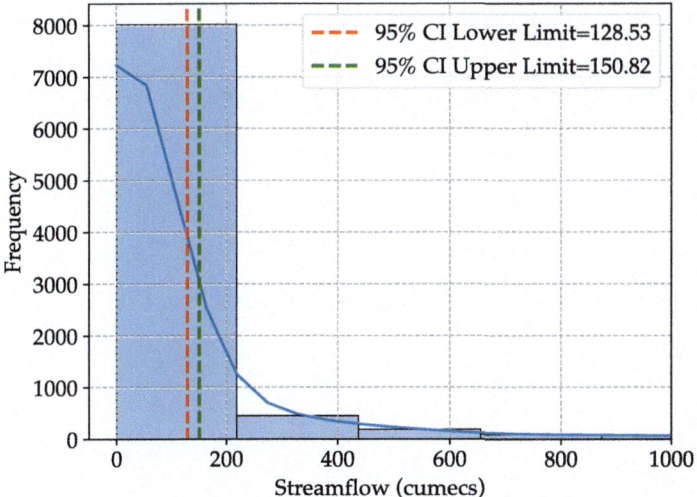

Fig. 9.3 Symmetric confidence interval estimate for mean

```python
1    """
2    Program for Symmetric confidence interval
3    estimate on 1D stream flow data
4    """
5    import pandas as pd
6    from scipy.stats import (norm, binom, tmean, tstd,
7                             t, nct, bootstrap, mstats)
8    import numpy as np
9
10   """
11   Load dataset
12   """
13   data = pd.read_csv(
14       filepath_or_buffer="../data/Godavari.csv",
15       sep=",",
16       header=0).dropna()
17   print("\nChecking data:")
18   try:
19       data["time"] = pd.to_datetime(
20           data['time'], infer_datetime_format=True)
21       print("   Date format is okay!\n")
22   except ValueError:
23       print("   Encountered error!\n")
24       pass
25   x = data[["Streamflow"]].to_numpy().squeeze()
26   del data
27   print("Read data file")
28
29   """
30   Symmetric confidence interval estimate
31   """
32   print("\n\nDoing symmetric CI estimate")
33   xbar = tmean(x)
34   xstd = tstd(x)
35   n = len(x)
36   qnt = t.ppf(q=[0.025, 0.975],
37               df=n - 1,
38               loc=0, scale=1)
39   print("{:0.4f} < x_mean < {:0.4f}".
40         format(xbar + qnt[0] * np.sqrt(xstd ** 2 / n),
41                xbar + qnt[1] * np.sqrt(xstd ** 2 / n))
42         )
43
44   print("\nDone!")
```

```
● ● ●            Terminal
>>> Checking data:
>>>     Date format is okay!
>>> Read data file
>>> Doing symmetric CI estimate
>>> 128.5336 < x_mean < 150.8192
>>> Done!
```

This program is written in Python and estimates a symmetric confidence interval for 1D stream flow data. The stream flow data is expected to be in a .csv file, with one of the columns labeled "Streamflow" representing the flow data and another column labeled "time" for timestamps. The file is assumed to be located in a data directory at the path "../data/Godavari.csv".

The first step in this program is to load the data from the .csv file into a Pandas dataframe, which makes it easy to manipulate tabular data in Python. The "dropna()" function removes any rows containing missing values.

Next, the program attempts to convert the "time" column to a standard datetime format for dealing with time series data. If the conversion is successful, the program prints "Date format is okay!" to the console. If the conversion fails due to the time data not being in the required format, it can be automatically recognized. In that case, the program catches the ValueError exception and prints "Encountered error!" to the console.

The Streamflow data is then extracted from the dataframe and converted into a one-dimensional NumPy array, a more efficient data structure for numerical computations. The dataframe is then deleted to clear memory.

Now, the program moves on to the main computation. It computes a symmetric confidence interval for the mean of the stream flow data. The sample mean and sample standard deviation of the data is first computed using the "tmean()" and "tstd()" functions from the SciPy stats module. The number of data points is also determined.

Next, the percent point function (also known as the inverse of the cumulative distribution function) of the t-distribution, "t.ppf()", is used to find the quantiles corresponding to the lower and the upper tails of the distribution. This is done at a significance level of 0.05 (hence the probabilities 0.025 and 0.975), which corresponds to a 95% confidence interval.

The lower and the upper limits of the confidence interval for the mean are then computed using these quantiles, the sample mean, the sample standard deviation, and the number of data points. These limits are printed to the console in the format ":0.4f < x_mean < :0.4f", where :0.4f is a placeholder for a floating point number with four digits after the decimal point. The confidence interval gives us an estimated range in which the true population mean is likely to lie with a 95% probability.

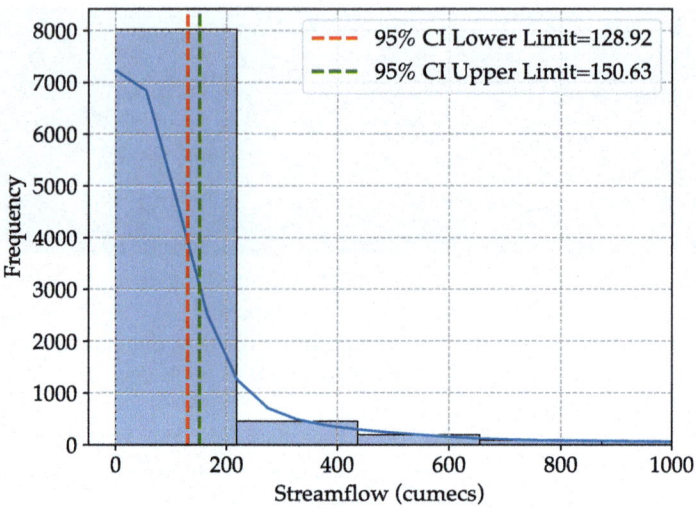

Fig. 9.4 Bootstrap confidence interval estimate

9.2.2.2 Bootstrap Confidence Interval Estimate

Bootstrap confidence interval estimate is a method to estimate the uncertainty of a statistic, such as the mean, using resampling techniques. The process involves repeatedly sampling observations from the original data, with replacement, calculating the mean each time, and then determining a range where a certain percentage of these means fall (Fig. 9.4).

```
1   """
2   Program for Bootstrap confidence
3   interval estimate on 1D stream flow data
4   """
5   import pandas as pd
6   from scipy.stats import (norm, binom, tmean, tstd,
7                             t, nct, bootstrap, mstats)
8   import numpy as np
9
10  """
11  Load dataset
12  """
13  data = pd.read_csv(
14      filepath_or_buffer="../data/Godavari.csv",
15      sep=",",
16      header=0).dropna()
17  print("\nChecking data:")
```

```
1   try:
2       data["time"] = pd.to_datetime(
3           data['time'], infer_datetime_format=True)
4       print("   Date format is okay!\n")
5   except ValueError:
6       print("   Encountered error!\n")
7       pass
8   x = data[["Streamflow"]].to_numpy().squeeze()
9   del data
10  print("Read data file")
11
12  """
13  Bootstrap confidence interval estimate
14  """
15  print("\n\nDoing Bootstrap CI estimate")
16  x = (x,)
17  ci = bootstrap(data=x,
18                  statistic=np.mean,
19                  confidence_level=0.95,
20                  n_resamples=2000,
21                  method="percentile")
22  lower_bound = ci.confidence_interval.low
23  upper_bound = ci.confidence_interval.high
24  print(f"Lower Bound: {lower_bound}")
25  print(f"Upper Bound: {upper_bound}")
26
27  print("\nDone!")
```

```
●  ●  ●        Terminal
>>> Checking data:
>>>     Date format is okay!
>>> Read data file
>>> Doing Bootstrap CI estimate
>>> Lower Bound: 128.921972747053
>>> Upper Bound: 150.630449446403
>>> Done!
```

The given Python program performs a Bootstrap confidence interval estimate on one-dimensional stream flow data. Bootstrap is a powerful statistical tool for estimating the accuracy and reliability of sample estimates, particularly in situations where the underlying distribution is unknown or difficult to handle analytically.

After loading the necessary libraries the program loads a dataset from a .csv file named "Godavari.csv" using Pandas' read_csv() function. It then drops any missing values (NaNs) found within the dataset. To ensure the date format in the data is consistent and understandable for further manipulation, the program attempts to convert the "time" column into a datetime object using Pandas' to_datetime()

function. Any errors during this conversion are caught and reported. Subsequently, it isolates the "Streamflow" column from the dataframe and converts it into a NumPy array for easier numerical manipulation, then deletes the original dataframe to free up memory.

Once the data is prepared, the main part of the program begins by performing the Bootstrap confidence interval estimation. The bootstrap() function from the scipy.stats library is used for this purpose. The function takes the stream flow data, the statistic of interest (in this case, the mean), the desired confidence level (95%), the number of resampling iterations (2000), and the method for estimating the confidence interval ("percentile") as inputs.

After executing the bootstrap function, the program extracts the lower and upper bounds of the confidence interval from the result and prints them. These bounds represent the range within which the true population mean is likely to lie with a 95% level of confidence, based on the bootstrap resampling of the given data.

9.2.3 For Quantiles

9.2.3.1 Non-parametric Confidence Interval Estimate

A non-parametric confidence interval estimate for quantiles involves determining the range within which a particular quantile of a population is likely to lie, without assuming that the population follows any specific distribution (Fig. 9.5).

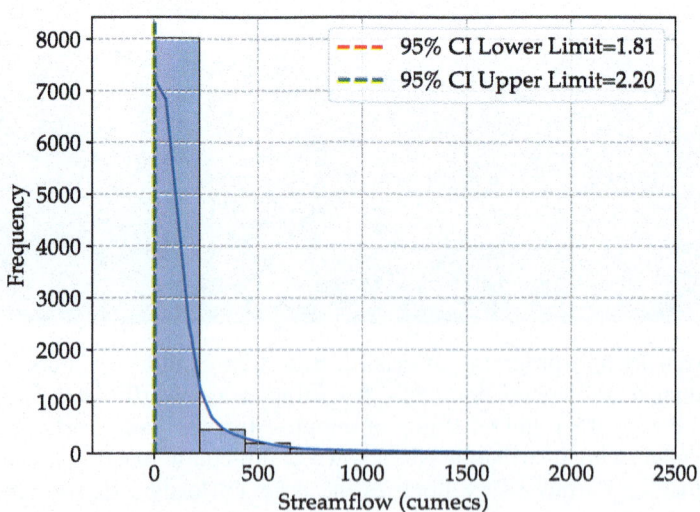

Fig. 9.5 Non-parametric confidence interval estimate for quantiles

```python
"""
Program for non-parametric confidence
interval estimate on 1D stream flow data
"""
import pandas as pd
from scipy.stats import (norm, binom, tmean, tstd,
                         t, nct, bootstrap, mstats)
import numpy as np


"""
Load dataset
"""
data = pd.read_csv(
    filepath_or_buffer="../data/Godavari.csv",
    sep=",",
    header=0).dropna()
print("\nChecking data:")
try:
    data["time"] = pd.to_datetime(
        data['time'], infer_datetime_format=True)
    print("   Date format is okay!\n")
except ValueError:
    print("   Encountered error!\n")
    pass
x = data[["Streamflow"]].to_numpy().squeeze()
del data
print("Read data file")



"""
Non-parametric confidence interval estimate on
percentile data
"""
print("\n\nDoing non-parametric CI estimate")
prob = 0.2
percentile = mstats.mquantiles(  # value of quantile
    a=x, prob=prob,
    alphap=0, betap=0
    )
```

```python
40  print("Quantile value =", percentile)
41
42  qnt = binom.ppf(q=[0.025, 0.975],    # position of quantiles
43                  n=len(x), p=prob,
44                  loc=0)
45  qnt = np.cast[int](qnt)
46  print("95% CI positions =", qnt)
47
48  result = binom.pmf(k=np.arange(qnt[0]-1, qnt[1]),
49                     n=len(x),
50                     p=prob,
51                     loc=0)
52  print("Probability sum upto 0.2 =", result.sum())
53
54
55  # Using bootstrap method
56  def percentile_statistic(data, axis=1):
57      percentile = mstats.mquantiles(   # value of quantile
58          a=data, prob=0.2,
59          alphap=0, betap=0, axis=axis
60      )
61      percentile = np.squeeze(percentile)
62      return percentile
63
64
65  x = (x,)
66  ci = bootstrap(data=x,
67                 statistic=percentile_statistic,
68                 confidence_level=0.95,
69                 n_resamples=10000,
70                 method="percentile")
71
72  # Get the lower and upper bounds of the confidence interval
73  lower_bound = ci.confidence_interval.low
74  upper_bound = ci.confidence_interval.high
75
76  print("\n\n{}th Percentile Bootstrap Confidence Interval:"
77        .format(prob*100))
78  print(f"Lower Bound: {lower_bound}")
79  print(f"Upper Bound: {upper_bound}")
80
81  print("\nDone!")
```

```
● ● ●            Terminal
>>> Checking data:
>>>    Date format is okay
>>> Read data file
>>> Doing non-parametric CI estimate
>>> Quantile value = [2.]
>>> 95% CI positions = [1741 1890]
>>> Probability sum upto 0.2 = 0.9508722528448205
>>> 20.0th Percentile Bootstrap Confidence Interval:
>>> Lower Bound: 1.8060000000000014
>>> Upper Bound: 2.2
>>> Done
```

The Python program demonstrated above performs a non-parametric estimation of the confidence interval on a one-dimensional stream flow dataset.

To begin, the Pandas and SciPy libraries are imported for data manipulation and statistical computations, respectively. The dataset is loaded from a .csv file titled "Godavari.csv". The "time" column is converted to a datetime format for temporal analyses.

After loading the main libraries, the data is extracted from the "Streamflow" column and converted to a NumPy array for subsequent computations. The original Pandas dataframe is deleted to clear memory, given that the focus of this analysis is on the stream flow data.

The next phase involves performing the non-parametric confidence interval estimation. The program first calculates the quantile value at a chosen probability (20% in this case). The positions for a 95% confidence interval (CI) are then determined. These positions represent the number of observations falling within the lower and the upper bounds of the CI.

The program then proceeds to compute the sum of the probabilities of the stream flow values that fall into the 95% CI. This operation essentially gives us the cumulative probability of the data falling within the CI at the 20% quantile.

The latter part of the program introduces the bootstrap method as an alternative approach to estimating the confidence interval. A new function, "percentile_statistic", is defined to calculate the 20% quantile value of the bootstrapped samples.

The bootstrap method, implemented through the "bootstrap" function, generates multiple resamples from the original data, applies the "percentile_statistic" function on each resample, and thereby calculates an array of bootstrapped quantile estimates. The number of resamples to generate is set to 10,000, while the confidence level is set to 95.

The program then prints the lower and the upper bounds of the bootstrapped 95% CI for the 20th percentile. These bounds tell us the range within which the true 20% quantile of the population is likely to fall with 95% confidence, based on our bootstrapped samples.

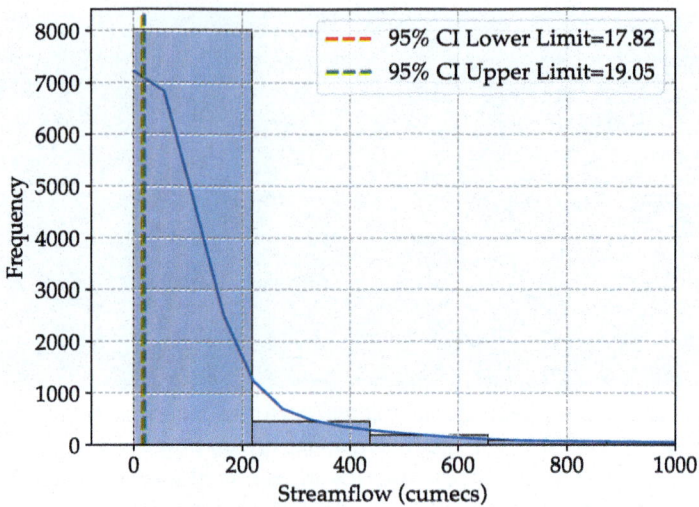

Fig. 9.6 Two-sided confidence interval estimate on log-normal streamflow data

9.2.3.2   Confidence Interval Estimate

A confidence interval estimate for quantiles provides a range within which a specific quantile of a population is likely to fall. It is a technique used in statistical inference, offering a probabilistic assessment of where the true value of the quantile lies based on a random sample from the population (Fig. 9.6).

```
1   """
2   Program for confidence interval estimate
3   on 1D stream flow lognormal data
4   """
5   import pandas as pd
6   from scipy.stats import (norm, binom, tmean, tstd,
7                            t, nct, bootstrap, mstats)
8   import numpy as np
9
10  """
11  Load dataset
12  """
13  data = pd.read_csv(
14      filepath_or_buffer="../data/Godavari.csv",
15      sep=",",
16      header=0).dropna()
17  print("\nChecking data:")
18  try:
19      data["time"] = pd.to_datetime(
20          data['time'], infer_datetime_format=True)
21      print("   Date format is okay!\n")
22  except ValueError:
23      print("   Encountered error!\n")
24      pass
```

```
27    x = data[["Streamflow"]].to_numpy().squeeze()
28    del data
29    print("Read data file")
30
31    """
32    Confidence interval estimate assuming
33    lognormal data
34    """
35
36    print("\n\nDoing CI estimate")
37    minx = np.min(x)
38    logx = np.log(x - minx + 1)
39    xbar = tmean(logx)
40    xstd = tstd(logx)
41    n = len(x)
42
43    qnt = norm.ppf(q=[0.9])
44    C90 = np.exp(xbar + qnt * xstd)
45
46    nc = -5 * qnt
47    qnt = nct.ppf(q=[0.05, 0.95],   # non-central t distribution
48                  df=n - 1,
49                  nc=nc,
50                  loc=0, scale=1)
51
52    print("{:.4f} < C90 < {:.4f}".
53          format(np.exp(xbar - 1 / np.sqrt(n) * qnt[1] * xstd),
54                 np.exp(xbar - 1 / np.sqrt(n) * qnt[0] * xstd)
55                 )
56          )
57    print("\nDone!")
```

```
●  ●  ●            Terminal
>>> Checking data:
>>>     Date format is okay
>>> Read data file
>>> Doing CI estimate
>>> 17.8163 < C90 < 19.0539
>>> Done
```

The program starts by importing necessary modules including Pandas for data manipulation, NumPy for numerical operations, and scipy.stats for various statistical calculations.

After loading the main libraries, the dataset is loaded using Pandas' read_csv() function, specifying the file path and the comma as the delimiter, and dropping any missing values. The loaded data is inspected and the "time" column is converted to a datetime object using Pandas' to_datetime function. If the conversion fails due to an incorrect format, the program catches the ValueError and displays an error message.

The "Streamflow" column from the data is then isolated, converted to a NumPy array, and squeezed to flatten it into one dimension. The original data variable is deleted to save memory.

Following this, the program calculates the confidence interval estimate, assuming the streamflow data follows a log-normal distribution. To work with a log-normal distribution, the program takes the natural logarithm of the streamflow values, after shifting them so that the minimum value becomes 1.

The mean (xbar) and standard deviation (xstd) of the log-transformed data are calculated, and the number of data points (n) is stored. The program then calculates the quantile (qnt) at 0.9 using the cumulative distribution function of the standard normal distribution (norm.ppf).

The 90% confidence level (C90) for the mean of the log-transformed data is then calculated using the inverse of the natural logarithm (np.exp). The program calculates the non-centrality parameter (nc), and computes the 5th and 95th percentiles of the non-central t-distribution (nct.ppf) using the calculated parameters.

Finally, the program prints the lower and the upper bounds of the 90% confidence interval for the mean of the original data (after converting back from the log-transformed scale). The result gives a range of values that contains the true mean of the streamflow data with a 90% probability.

9.3   Prediction Intervals

9.3.1   Non-parametric Prediction Interval

A non-parametric prediction interval for quantiles is a method of estimating the range where future observations are likely to fall, without making assumptions about the underlying distribution. It is particularly useful when data does not follow a known or easily identifiable distribution, relying instead on the actual data structure (Fig. 9.7).

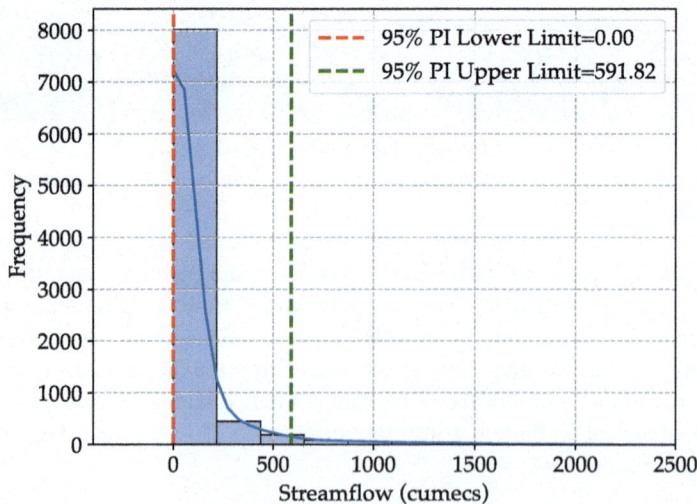

Fig. 9.7 Non-parametric prediction interval

```
1    """
2    Program for Non-parametric prediction
3    interval estimate on 1D stream flow data
4    """
5    import pandas as pd
6    from scipy.stats import (norm, binom, tmean, tstd,
7                             t, nct, bootstrap, mstats)
8    import numpy as np
9
10   """
11   Load dataset
12   """
13   data = pd.read_csv(
14       filepath_or_buffer="../data/Godavari.csv",
15       sep=",",
16       header=0).dropna()
17   print("\nChecking data:")
18   try:
19       data["time"] = pd.to_datetime(
20           data['time'], infer_datetime_format=True)
21       print("    Date format is okay!\n")
22   except ValueError:
23       print("    Encountered error!\n")
24       pass
25   x = data[["Streamflow"]].to_numpy().squeeze()
26   del data
27   print("Read data file")
28
29
30   """
31   Non-parametric prediction interval
32   """
33   print("\n\nDoing non-parametric PI estimate")
34   qi = mstats.mquantiles(
35       a=x, prob=[0.05, 0.95],
36       alphap=0, betap=0    # type 6 in R
37       )
38   print(f"Lower Bound: {qi[0]}")
39   print(f"Upper Bound: {qi[1]}")
40
41   print("\nDone!")
```

```
●  ●  ●          Terminal
>>> Checking data:
>>>     Date format is okay▌
>>> Read data file
>>> Doing non-parametric PI estimate
>>> Lower Bound: 0.0
>>> Upper Bound: 591.82
>>> Done▌
```

This Python program estimates a non-parametric prediction interval on a 1D streamflow dataset. The script utilizes the Pandas, NumPy, and SciPy libraries, powerful tools for data handling, mathematical operations, and statistical analysis.

The program begins by loading the dataset using the Pandas function "read_csv()" to read in a .csv file named "Godavari.csv" in the "data" directory. The "dropna()" method drops any missing data points from the dataset. To verify if the "time" column in the dataset is correctly formatted as datetime, the "to_datetime()" function from Pandas is used. The column named "Streamflow" is then converted to a NumPy array for further processing, and the original dataframe is deleted to conserve memory.

Following the loading and preprocessing of the data, the program proceeds to the main task: the non-parametric prediction interval estimate. This task is performed using the "mquantiles()" function from the "mstats" module of the SciPy library. The function calculates the stream flow data's quantiles (in this case, the 5th and 95th percentiles). These quantiles serve as the prediction interval, giving the range within which future observations are predicted to fall with a certain level of confidence (in this case, 90%). The "mquantiles()" function is called with the "alphap" and "betap" parameters both set to 0, which specifies that it should use the 6th type of quantile estimation method as used in the statistical software "R". The lower and upper bounds of the prediction interval are then printed to the console.

9.3.2   One-Sided Non-parametric Prediction Interval

A one-sided non-parametric prediction interval for quantiles is a statistical estimate that provides a lower or upper bound, but not both, within which a future observation will fall with a certain probability. It relies on the empirical distribution of the data, rather than making assumptions about the underlying population distribution (Fig. 9.8).

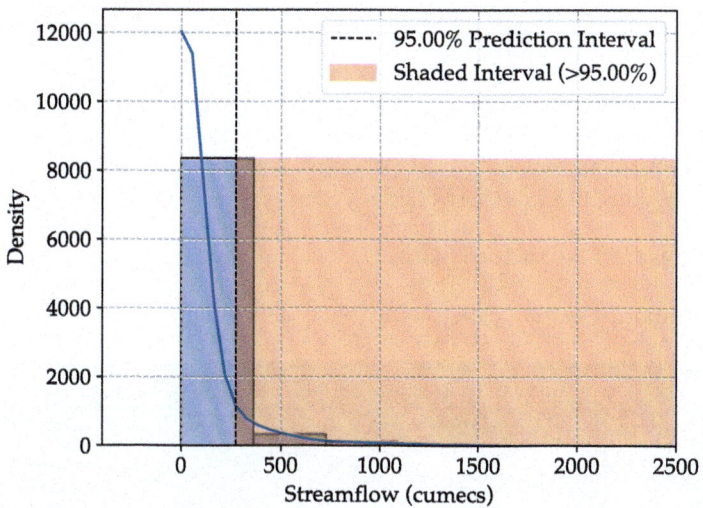

Fig. 9.8 One-sided non-parametric prediction interval estimate

```
1   """
2   Program for One-side Non-parametric prediction
3   interval estimate on 1D stream flow data
4   """
5   import pandas as pd
6   from scipy.stats import (norm, binom, tmean, tstd,
7                            t, nct, bootstrap, mstats)
8   import numpy as np
9
10  """
11  Load dataset
12  """
13  data = pd.read_csv(
14      filepath_or_buffer="../data/Godavari.csv",
15      sep=",",
16      header=0).dropna()
17  print("\nChecking data:")
18  try:
19      data["time"] = pd.to_datetime(
20          data['time'], infer_datetime_format=True)
21      print("    Date format is okay!\n")
22  except ValueError:
23      print("    Encountered error!\n")
24      pass
25  x = data[["Streamflow"]].to_numpy().squeeze()
26  del data
27  print("Read data file")
```

```python
1   """
2   Non-parametric prediction interval
3   """
4   print("\n\nDoing non-parametric PI estimate")
5   qi = mstats.mquantiles(
6       a=x, prob=0.9,
7       alphap=0, betap=0   # type 6 in R
8       )
9   print(f"Quantile value: {qi}")
10
11  print("\nDone!")
```

```
● ● ●              Terminal

>>> Checking data:
>>>     Date format is okay!
>>> Read data file
>>> Doing non-parametric PI estimate
>>> Quantile value: [274.888]
>>> Done!
```

The above Python program performs a one-sided non-parametric prediction interval estimate on a one-dimensional stream flow dataset. The data for this analysis is taken from a .csv file named "Godavari.csv".

The program first imports the necessary modules. These include Pandas for data manipulation, scipy.stats for statistical analysis, and NumPy for numerical computation.

After importing the necessary modules, the program reads the dataset using the Pandas' read_csv() function. The .csv file is located at the path "../data/Godavari.csv". The program removes any missing values in the data using the dropna() function.

After loading the dataset, the program checks whether the "time" column can be converted to datetime format using Pandas to_datetime() function.

Then, the program extracts the "Streamflow" column from the dataframe and converts it into a NumPy array using the to_numpy() function. The squeeze() function is used to ensure that the resulting array has one dimension.

Next, the program performs the non-parametric prediction interval estimate. The function mstats.mquantiles() from the scipy.stats module is used for this purpose. This function computes the quantile of the streamflow data at a probability level of 0.9, equivalent to the 90th percentile. The arguments alphap=0 and betap=0 specify the estimation method for the quantiles (in this case, a method equivalent to type 6 in R is chosen). The calculated quantile value is then printed to the console.

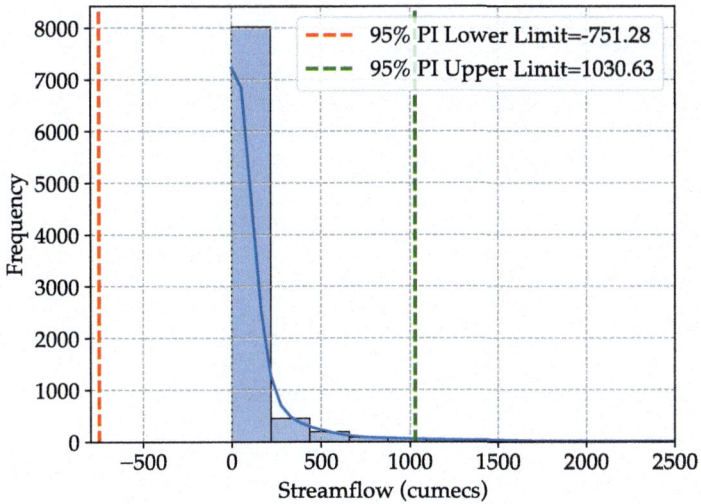

Fig. 9.9 Two-sided parametric prediction interval estimate

9.3.3 Two-Sided Parametric Prediction Interval

A two-sided parametric prediction interval for quantiles provides a range in which future observations are expected to fall with a certain probability. This approach does not assume any specific statistical distribution for the data. It is useful when the underlying distribution is unknown or does not adhere to standard parametric forms (Fig. 9.9).

```python
"""
Program for Two-sided parametric prediction
interval estimate on 1D stream flow data
"""
import pandas as pd
from scipy.stats import (norm, binom, tmean, tstd,
                         t, nct, bootstrap, mstats)
import numpy as np


"""
Load dataset
"""
data = pd.read_csv(
    filepath_or_buffer="../data/Godavari.csv",
    sep=",",
    header=0).dropna()
print("\nChecking data:")
```

```
1   try:
2       data["time"] = pd.to_datetime(
3           data['time'], infer_datetime_format=True)
4       print("   Date format is okay!\n")
5   except ValueError:
6       print("   Encountered error!\n")
7       pass
8   x = data[["Streamflow"]].to_numpy().squeeze()
9   del data
10  print("Read data file")
11
12
13  """
14  Two-sided parametric prediction interval
15  """
16  print("\n\nDoing two-sided parametric PI estimate")
17  xbar = tmean(x)
18  xstd = tstd(x)
19  n = len(x)
20  qnt = t.ppf(q=[0.05, 0.95],
21              df=n-1,
22              loc=0, scale=1)
23  print("{:.4f} < x_mean < {:.4f}".
24        format(xbar+qnt[0]*np.sqrt(xstd**2 + xstd**2/n),
25               xbar+qnt[1]*np.sqrt(xstd**2 + xstd**2/n))
26        )
27
28  print("\nDone!")
```

```
●●●        Terminal
>>> Checking data:
>>>     Date format is okay!
>>> Read data file
>>> Doing two-sided parametric PI estimate
>>> -751.2762 < x_mean < 1030.6290
>>> Done!
```

The given Python program computes a two-sided parametric prediction interval for a dataset of 1D stream flow data. This type of prediction interval provides a range within which future observations are likely to fall, with a certain level of confidence.

The script starts by importing necessary Python packages, including Pandas for data manipulation, and various functions from the scipy.stats library for statistical computations.

After importing the necessary modules, the program loads the dataset, which is expected to be a .csv file. It reads the data file using the Pandas function "pd.read_csv()". It is designed to handle missing data by dropping rows where data is not available, using the "dropna()" function. It then attempts to convert the "time" column into a datetime format, which can be particularly useful for time series analysis.

After the initial preprocessing, the program extracts the "Streamflow" column data and converts it into a NumPy array. The extracted stream flow data is assigned to the variable "x", and the original dataset is deleted to save memory.

The main computation of the script lies in the section titled "Two-sided parametric prediction interval". Here, the program first calculates the mean ("xbar") and standard deviation ("xstd") of the streamflow data. The size of the data, "n", is also determined.

The program then computes the quantiles at 0.05 and 0.95 (representing the lower and upper bounds of the 90% prediction interval) of the Student's t-distribution with "n-1" degrees of freedom. These quantiles are used to calculate the lower and upper limits of the two-sided prediction interval.

Finally, the computed prediction interval is printed to the console. This range represents the values within which future observations are expected to fall with a 90% confidence level, based on the provided streamflow data.

9.3.4 Asymmetric Prediction Interval

An asymmetric prediction interval for quantiles provides a range within which a future observation is expected to fall with a certain probability. Unlike symmetric intervals, these intervals adjust for skewness in the data, meaning the distance to the lower and upper bounds can vary, better reflecting the underlying data distribution (Fig. 9.10).

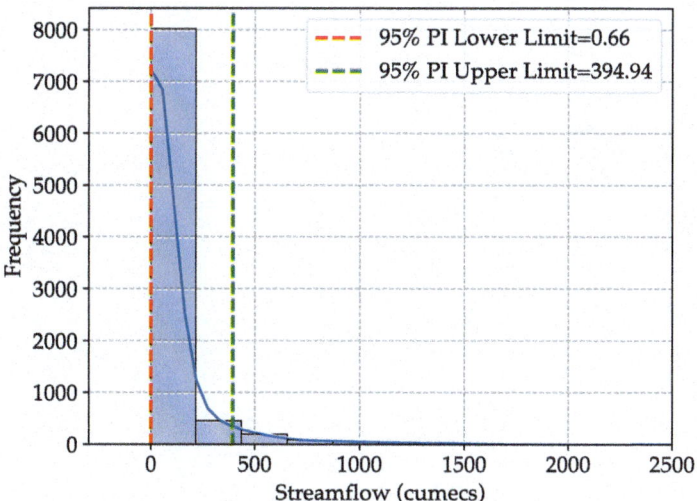

Fig. 9.10 Asymmetric prediction interval estimate

```python
"""
Program for asymmetric prediction
interval estimate on 1D stream flow data
"""
import pandas as pd
from scipy.stats import (norm, binom, tmean, tstd,
                         t, nct, bootstrap, mstats)
import numpy as np


"""
Load dataset
"""
data = pd.read_csv(
    filepath_or_buffer="../data/Godavari.csv",
    sep=",",
    header=0).dropna()
print("\nChecking data:")
try:
    data["time"] = pd.to_datetime(
        data['time'], infer_datetime_format=True)
    print("   Date format is okay!\n")
except ValueError:
    print("   Encountered error!\n")
    pass
x = data[["Streamflow"]].to_numpy().squeeze()
del data
print("Read data file")


"""
Asymmetric prediction interval
"""
print("\n\nDoing asymmetric PI estimate")
minx = np.min(x)
logx = np.log(x - minx + 1)
xbar = tmean(logx)
xstd = tstd(logx)
n = len(x)
qnt = t.ppf(q=[0.05, 0.95],
            df=n - 1,
            loc=0, scale=1)
print("{:.4f} < x_mean < {:.4f}".format(
    np.exp(xbar + qnt[0] * np.sqrt(xstd ** 2 + xstd ** 2 / n)),
    np.exp(xbar + qnt[1] * np.sqrt(xstd ** 2 + xstd ** 2 / n))
)
)

print("Done!")
```

```
● ● ●            Terminal
>>> Checking data:
>>>    Date format is okay█
>>> Read data file
>>> Doing asymmetric PI estimate
>>> 0.6619 < x_mean < 394.9448
>>> Done█
```

This Python program performs an asymmetric prediction interval estimate on a one-dimensional stream flow dataset. The data for this exercise is stored in a .csv file named "Godavari.csv".

First, the necessary modules are imported. These include Pandas for data handling, SciPy's stats module for statistical functions, and NumPy for numerical operations.

Next, the program loads the dataset using the Pandas "read_csv" function. The data is then checked to ensure that the "time" column, which should contain datetime information, is properly formatted. If there is a ValueError during conversion to Pandas datetime format, it implies the date format is not recognized and an error message is printed. If no error is encountered, a message confirming the same is printed.

The "Streamflow" column of the data is then extracted and converted to a one-dimensional NumPy array for further analysis, and the original dataframe is deleted.

The program then proceeds to perform the main task: calculating the asymmetric prediction interval. It begins by transforming the data to a logarithmic scale to handle potential skewness in the data, ensuring all values are positive by subtracting the minimum value of the dataset and adding 1 before taking the logarithm.

Next, the mean and standard deviation of the transformed data are computed. Then, the percentiles corresponding to the 5th and 95th percentiles of a Student's t-distribution are calculated. These percentiles serve as the lower and the upper bounds of the 90% prediction interval.

Finally, the program prints the estimated asymmetric prediction interval. The interval is converted back to the original scale (from the log scale) using the exponential function. The interval gives the range in which we expect a new observation to fall, with 90% confidence, assuming that the new observation is from the same population as the data.

9.4 Quantile Regression

Quantile regression is a statistical technique aimed at estimating and conducting inferences about conditional quantile functions. Unlike standard regression models that predict mean outcomes, quantile regression provides a complete picture of the possible outcome values at different points of the distribution (like median, quartiles), thus offering a more robust and comprehensive analysis.

```python
"""
Program to do Quantile regression
on synthetic dataset.
"""

from sklearn.linear_model import QuantileRegressor
import numpy as np
import matplotlib.pyplot as plt


"""
Generate synthetic normal data
"""
np.random.seed(21)
x = np.arange(start=0.5 * np.pi, stop=2 * np.pi, step=0.05)
X = x.reshape(x.shape[0], -1)
mean_ = 10.0 + 0.1 * np.sin(x)
scale_ = 0.02 + 0.02 * np.abs(x)
y_true = np.random.normal(loc=mean_,
                          scale=scale_,
                          size=x.shape[0])


"""
Quantile regression
"""
idx_up = 0
idx_dn = 0
x_good = list()
y_good = list()

preds = dict()
x_outlier = list()
y_outlier = list()

qnt = [0.1, 0.25, 0.5, 0.75, 0.9]

for q in qnt:
    qreg = QuantileRegressor(quantile=q,
                             alpha=0,
                             solver='highs')
    qreg.fit(X=X, y=y_true)
    preds[q] = qreg.predict(X)

    # Segregate the detected outliers using index
    if q == qnt[0]:
        idx_dn = preds[q] > y_true
```

```
37      elif q == qnt[-1]:
38          idx_up = preds[q] < y_true
39
40  idx_outlier = np.logical_or(idx_up, idx_dn)
41  x_outlier.append(x[idx_outlier])
42  y_outlier.append(y_true[idx_outlier])
43  x_outlier = np.array(x_outlier).ravel()
44  y_outlier = np.array(y_outlier).ravel()
45
46  idx_good = ~np.logical_or(idx_up, idx_dn)
47  x_good.append(x[idx_good])
48  y_good.append(y_true[idx_good])
49  x_good = np.array(x_good).ravel()
50  y_good = np.array(y_good).ravel()
51
52  """
53  Plot the scatter and then show
54  fit of the data on the scatter
55  """
56  plt.figure(figsize=[8, 6])
57  # Quantile plots
58  plt.plot(X, y_true,
59          color="black",
60          linestyle="dashed",
61          label="True y")
62  for q, y_pred in preds.items():
63      plt.plot(X, y_pred, label="Q={}".format(q))
64
65  # Outliers data plot
66
67  plt.scatter(x=x_outlier, y=y_outlier, marker='o',
68              alpha=0.5, color='k',
69              label='Outside: Q={} and Q={}'.format(qnt[0],
70                                                     qnt[-1]))
71
72  # Inside data plot
73  plt.scatter(x=x_good, y=y_good, marker='x',
74              alpha=0.5, color='b',
75              label='Within: Q={} and Q={}'.format(qnt[0],
76                                                    qnt[-1]))
```

```
77  plt.xlabel('x')
78  plt.ylabel('y')
79  plt.ylim([9.5, 10.8])
80  plt.grid(ls='--')
81  plt.legend(loc='upper center', ncol=3)
82  plt.show()
83  plt.tight_layout()
84  plt.savefig('quantile_reg.pdf', dpi=300)
```

The above Python script demonstrates the implementation of Quantile Regression on a synthetic dataset. The script is divided into several sections to systematically generate the synthetic data, apply quantile regression, identify outliers, and visualize the results.

In the first section, a synthetic dataset is generated using a random seed. The "x" values range from half of pi to double pi with an interval of 0.05. The "y_true" values are generated by sampling from a normal distribution with mean as "mean_" and standard deviation as "scale_". The mean and standard deviation are dependent on "x", creating heteroscedastic data (i.e., data with non-constant variance).

The script then proceeds to apply quantile regression. Five quantile levels (10%, 25%, 50%, 75%, and 90%) are considered. For each quantile, a QuantileRegressor model from the sklearn library is created and fitted to the synthetic data. The predictions of these models for each quantile level are stored in the dictionary "preds".

The program then identifies outliers in the data. Observations are considered outliers if they lie outside the region defined by the 10% and 90% quantile regressions. The indices of these outliers are determined, and the corresponding "x" and "y_true" values are stored in "x_outlier" and "y_outlier", respectively. Simultaneously, "x" and "y_true" values that lie within the specified quantiles are stored in "x_good" and "y_good", respectively.

The last section of the script involves visualizing the results (Fig. 9.11). It generates a scatter plot with different markings for outliers and non-outliers, overlaid with lines representing the true "y" values and the quantile regressions. The plot is customized with labels, grid lines, and a legend for better readability. The figure is saved as "quantile_reg.pdf" in the current directory at a resolution of 300 dpi.

9.5 Maximum Likelihood Estimation (MLE)

Maximum Likelihood Estimation (MLE) is a statistical method that estimates the parameters of a model by finding the parameter values that maximize the likelihood function. The resulting estimates maximize the probability of observing the data given in the parameter values, assuming the chosen model is correct.

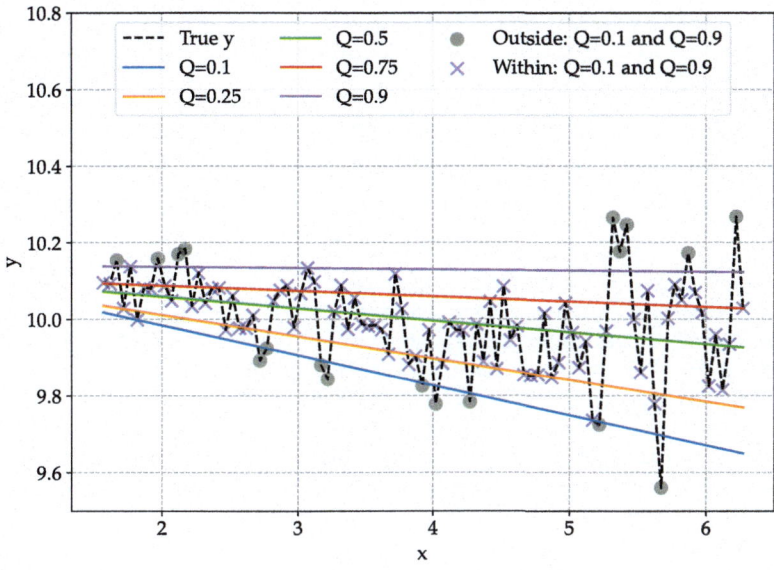

Fig. 9.11 Quantile regression plot to detect outliers beyond the specified quantiles, i.e., 10% and 90%

```
1   """
2   Demonstration of Maximum-likelihood estimation
3   for log-normal data
4   """
5   from scipy.stats import norm
6   import numpy as np
7   import matplotlib.pyplot as plt
8   import pandas as pd
9
10
11  np.random.seed(11)
12  plt.rc('text', usetex=True)
13
14
15  """
16  Loading the dataset
17  """
18  data = pd.read_csv(
19      filepath_or_buffer="../data/Godavari.csv",
20      sep=",",
21      header=0
22  ).dropna()
23  streamflow_data = np.array(data["Streamflow"])
24  streamflow_data = streamflow_data[streamflow_data > 0]
25
26  # Log transformation of stream data
```

```
27  data = np.log10(streamflow_data)
28
29  # Plot preliminary histogram
30  plt.figure()
31  plt.hist(data,
32          density=True, bins=100, log=True,
33          label="log(data)")
34
35  """
36  Fit the normal distribution parameters.
37  These formulae can be used to get the
38  parameters of the lognormal distribution:
39  mu_ = ln(mu) - (1/2) * ln(1 + (sig^2 / mu^2))
40  sig_ = (ln(1 + (sig^2 / mu^2)))**0.5
41  """
42  fit = norm.fit(data=data)   # fit Normal distribution (MLE)
43  mu, scale = fit   # in Normal distribution
44
45  N = int(data.shape[0]/2)   # generate N samples
46  samples = norm.rvs(loc=mu, scale=scale, size=N)
47
48  plt.hist(samples,   # Plot MLE estimated distribution
49          density=True, bins=100, log=True,
50          label="log(data$_{" + "MLE" + "}$)",
51          alpha=.6)
52  plt.xlabel("Streamflow (m$^3$/s)")
53  plt.ylabel("log(Frequency)")
54  plt.legend()
55  plt.grid(ls="--")
56  plt.tight_layout()
57  plt.savefig("streamflow_log_hist.pdf", dpi=300)
```

This Python script demonstrates the maximum likelihood estimation (MLE) process for log-normal data using a dataset of streamflow measurements. It utilizes the "scipy.stats" and "NumPy" libraries for statistical and mathematical operations, "matplotlib.pyplot" for plotting, and "Pandas" for data handling.

The script starts by setting a random seed for reproducibility and enabling the use of LaTeX for text formatting in matplotlib. It then loads a dataset from a .csv file located in the given path. After loading, it eliminates null values from the dataset and isolates the "Streamflow" data. It retains only positive streamflow data and applies a log10 transformation to it.

Following the data preparation, the script visualizes the log-transformed data by plotting a histogram, which shows the log of the frequency of streamflow data against streamflow in cubic meters per second (m^3/s).

After the preliminary visualization, the script proceeds to fit a normal distribution to the log-transformed streamflow data using the fit function from the "scipy.stats" library's "norm" module, which employs the method of maximum likelihood esti-

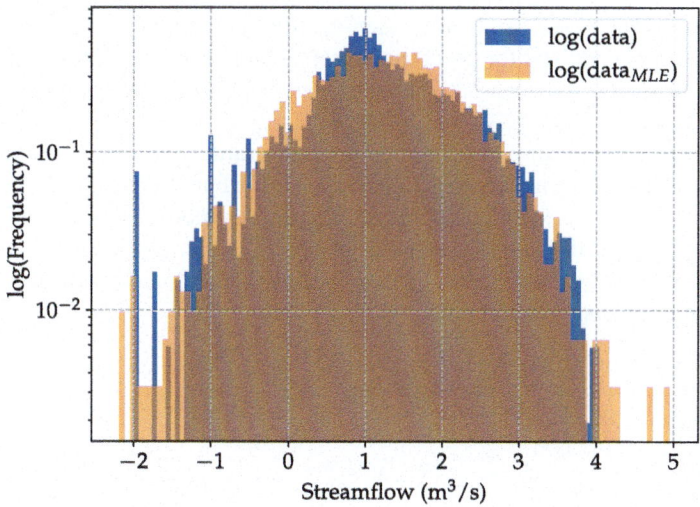

Fig. 9.12 Logarithm transform of the streamflow data

mation (MLE). The fit function returns the estimated parameters of the normal distribution: the mean ("mu") and the standard deviation ("scale").

The script then generates a set of random samples from the fitted normal distribution, with the number of samples ("N") being half the size of the original dataset.

Finally, the script creates another histogram of the generated samples, superimposing it on the original histogram (Fig. 9.12). This second histogram represents the fitted normal distribution based on the maximum likelihood estimation. The script saves the final figure as a PDF file. By comparing the two histograms in the figure, one can visually assess the goodness-of-fit of the MLE-based normal distribution to the original log-transformed streamflow data.

9.6 Monte Carlo Uncertainty Propagation

Monte Carlo uncertainty propagation is an effective technique for quantifying and managing the uncertainty inherent in mathematical models, including those used in hydrology, engineering, finance, and other fields. This method uses random sampling to evaluate model output uncertainty based on model inputs' uncertainties.

The key concept behind Monte Carlo uncertainty propagation is that uncertainty in the inputs of a model leads to uncertainty in the output of the model. If we can represent the uncertainty in the inputs as probability distributions, we can use these distributions to generate multiple sets of input data. We can then run the model with each of these input sets and observe the resulting distribution of outputs. This output distribution estimates the uncertainty in the model's predictions.

Monte Carlo methods are handy when dealing with complex, nonlinear models where analytical solutions for uncertainty propagation are challenging or impossible to derive. However, they can be computationally intensive, especially for complex models or when a high precision is required.

Monte Carlo uncertainty propagation provides valuable insights into the behavior of a system under uncertainty, allowing competent decision-making. Exploring a wide range of possible outcomes helps identify potential risks and opportunities, supporting risk management, design optimization, and scenario analysis. It has become an indispensable tool requiring sophisticated uncertainty analysis and decision-making under uncertainty.

```
"""
Monte Carlo error propagation
on a simple forward model:
Manning's equation
V = (1/n) * R^(2/3) * S^(1/2),
where,
V is the flow velocity,
n is the Manning's roughness coefficient,
R is the hydraulic radius, and
S is the channel slope
"""

import numpy as np
from scipy import stats
import matplotlib.pyplot as plt

plt.rc('text', usetex=True)
np.random.seed(11)


def compute_velocity(n=1., b=1., h=1., S=1.):
    R = (b * h) / (2 * b + h)
    V = (1 / n) * R ** (2 / 3) * S ** (1 / 2)
    return V



"""
Monte-Carlo simulation #1
"""
Nsim = 100000
n_var = stats.halfnorm.rvs(loc=0.030, scale=1e-2, size=Nsim)
b_var = stats.norm.rvs(loc=2.0, scale=1e-1, size=Nsim)
h_var = stats.norm.rvs(loc=1.3, scale=1e-1, size=Nsim)
S_var = stats.norm.rvs(loc=np.tan(15 * np.pi / 180),
                       scale=1e-2, size=Nsim)
```

```python
36   data = np.vstack([n_var, b_var, h_var, S_var]).T
37
38   V_var = np.zeros_like(n_var)
39   for i in range(Nsim):
40       V_var[i] = compute_velocity(
41           n=data[i, 0],
42           b=data[i, 1],
43           h=data[i, 2],
44           S=data[i, 3]
45       )
46
47
48   def normal_distribution(x, mu, sig):
49       rv = 1.0 / (sig * np.sqrt(2.0 * np.pi))
50       normal_values = rv * np.exp(
51           -(x - mu) ** 2.0 / (2.0 * sig ** 2)
52       )
53       return normal_values
54
55
56   # Plots
57   xmin, xmax = V_var.min(), V_var.max()
58   x = np.linspace(xmin, xmax, Nsim)
59   y = normal_distribution(x, np.mean(V_var), np.std(V_var))
60
61   plt.hist(V_var, density=True, bins=150,
62            label="$V_{" + "MC" + "}$")
63   plt.plot(x, y, color='red', lw=3, label="Normal")
64   plt.xlabel("Velocity (m/s)")
65   plt.ylabel("Relative Probability")
66   plt.xlim(xmin, xmax)
67   plt.legend()
68   plt.grid(ls="--")
69   plt.tight_layout()
70   plt.savefig("mc1.pdf", dpi=300)
71
72   """
73   Monte-Carlo simulation #2
74   """
75   Nsim = 100000
76   n_var = stats.halfnorm.rvs(loc=0.030, scale=1e-1, size=Nsim)
77   b_var = stats.norm.rvs(loc=2.0, scale=1e-1, size=Nsim)
```

```
78   h_var = stats.norm.rvs(loc=1.3, scale=1e-1, size=Nsim)
79   S_var = stats.norm.rvs(loc=np.tan(15 * np.pi / 180),
80                          scale=1e-2, size=Nsim)
81   data = np.vstack([n_var, b_var, h_var, S_var]).T
82
83   V_var = np.zeros_like(n_var)
84   for i in range(Nsim):
85       V_var[i] = compute_velocity(
86           n=data[i, 0],
87           b=data[i, 1],
88           h=data[i, 2],
89           S=data[i, 3])
90
91   # Plots
92   xmin, xmax = V_var.min(), V_var.max()
93   x = np.linspace(xmin, xmax, Nsim)
94   y = normal_distribution(x, np.mean(V_var), np.std(V_var))
95
96   plt.figure()
97   plt.hist(V_var, density=True, bins=150,
98           label="$V_{" + "MC" + "}$")
99   plt.plot(x, y, color='red', lw=3, label="Normal")
100  plt.xlabel("Velocity (m/s)")
101  plt.ylabel("Relative Probability")
102  plt.xlim(xmin, xmax)
103  plt.legend()
104  plt.grid(ls="--")
105  plt.tight_layout()
106  plt.savefig("mc2.pdf", dpi=300)
```

The above program applies Monte Carlo simulation to propagate errors in Manning's equation, a commonly used formula in fluid dynamics to estimate the velocity of flow in an open channel.

The program begins by defining Manning's equation as a function, where the flow velocity V is computed based on Manning's roughness coefficient (n), the hydraulic radius (R), and the channel slope (S).

Then, the Monte Carlo simulation is conducted in two parts, each with 100,000 iterations. In each simulation, random values for the parameters n, b (width), h (height), and S are generated following specific statistical distributions. The half-normal distribution is used for the roughness coefficient, while normal distributions are used for the width, height, and channel slope.

For each set of parameters generated, the velocity V is calculated using the function defined earlier, and the results are stored in an array. This process is repeated for each iteration of the simulation, generating a distribution of possible velocities based on the variations in the input parameters.

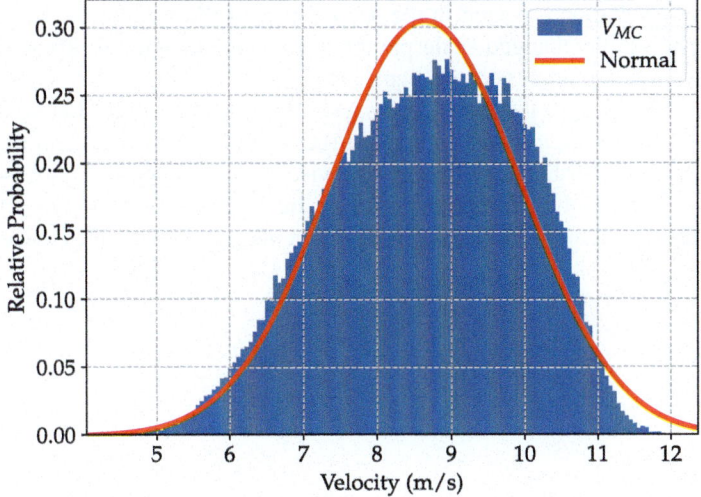

Fig. 9.13 Monte Carlo estimate for roughness coefficient = 0.01

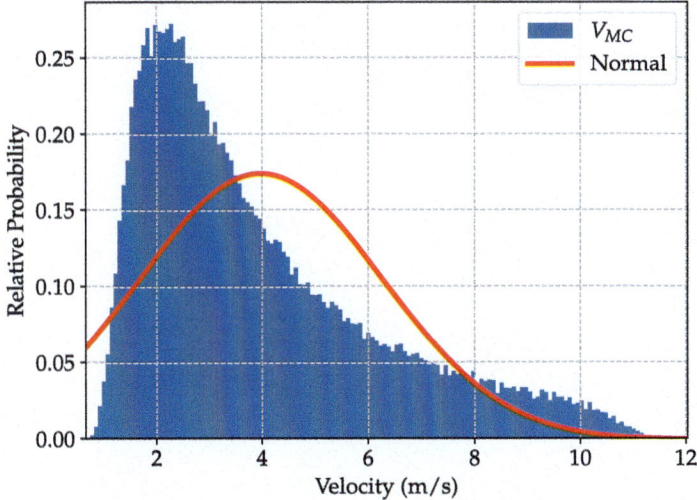

Fig. 9.14 Monte Carlo estimate with increased roughness coefficient = 0.1

Once the velocities have been calculated for all iterations, the program constructs a histogram of the simulated velocities and overlays it on a normal distribution with the mean and standard deviation of the simulated velocities. A single plot helps to compare the distribution of the velocities obtained from the Monte Carlo simulation with a normal distribution (Fig. 9.13).

The process is then repeated in the second part of the simulation, with the only difference being the changed scale parameter of the half-normal distribution used for generating the roughness coefficient.

Finally, the program generates and saves two graphs, one for each part of the simulation, illustrating the distributions of the simulated velocities and their corresponding normal distributions. This gives a visual representation (Fig. 9.14) of how variations in the input parameters can affect the velocity calculated by Manning's equation, thereby demonstrating the propagation of errors in a simple forward model.

Part III
Surface and Subsurface Water

Chapter 10
Introduction

In fluid dynamics, the Navier-Stokes equations along with some initial conditions are used to describe the velocity vector field of a fluid. The equations are a consequence of Newton's second law of motion, the stress in a fluid due to viscosity, and the applied pressure. Most of the real flow phenomena can be addressed by solving the nonlinear set of partial differential equations. This requires advanced nonlinear solvers which although when used do not guarantee convergence. However, we often use simplified versions of these equations for our own applications. This essentially means that the original nonlinear equations are reduced to linear forms by making some assumptions, like in the case of one-dimensional flow.

The general form of the Navier-Stokes equations can be given by the following equation:

$$\rho \frac{D\mathbf{v}}{Dt} = -\nabla p + \nabla \cdot \sigma_T + \mathbf{f} \qquad (10.1)$$

This equation is derived by the application of the continuity equation to the mass and momentum conservation equations. A few of the concepts that are helpful in understanding the terms dictated in the Eq. 10.1 are described now.

Material derivative is a required concept for the momentum conservation equation. We define the material derivative as the rate of change of an intensive property, u, in the velocity field, $\mathbf{v}$. Mathematically, it can be expressed as

$$\frac{Du}{Dt} = \frac{du}{dt} + \mathbf{v} \cdot \nabla u \qquad (10.2)$$

The second term in the Eq. 10.2 represents the directional derivative of u along the direction of the velocity field, $\mathbf{v}$. Newton's second law can be stated as

© The Author(s), under exclusive license to Springer Nature Singapore Pte Ltd. 2024
A. Kumar and M. Saharia, *Python for Water and Environment*, Innovations in Sustainable Technologies and Computing, https://doi.org/10.1007/978-981-99-9408-3_10

$$b = \rho \frac{d}{dt} v(x, y, z, t) \tag{10.3}$$

$$b = \rho \left(\frac{\partial v}{\partial t} + \frac{\partial v}{\partial x} \frac{\partial x}{\partial t} + \frac{\partial v}{\partial y} \frac{\partial y}{\partial t} + \frac{\partial v}{\partial z} \frac{\partial z}{\partial t} \right) \tag{10.4}$$

If velocity, v, is the intensive property, then Eq. 10.3 can be rewritten as

$$b = \rho \left(\frac{\partial v}{\partial t} + v \cdot \nabla v \right) \tag{10.5}$$

or, $$b = \rho \frac{Dv}{Dt}, \qquad \text{(using Eq. 10.2)} \tag{10.6}$$

Further assuming that the total body force on the infinitesimal fluid cubes is due to stresses within the fluid and external forces, we can state that:

$$b = \nabla \cdot \sigma + f \tag{10.7}$$

Here, f represents the external forces, $\nabla \cdot \sigma$ represents the divergence of the stress tensor and b is the body force. The stress tensor can be written as

$$\sigma = \begin{pmatrix} \sigma_{xx} & \sigma_{xy} & \sigma_{xz} \\ \sigma_{yx} & \sigma_{yy} & \sigma_{yz} \\ \sigma_{zx} & \sigma_{zy} & \sigma_{zz} \end{pmatrix} \tag{10.8}$$

The stress tensor, σ, can be further composed of two components—the volumetric part and the deviatoric part. The former is responsible for change in the body while the latter takes care of the change in shape or the deformation. Mathematically, it can be written as

$$\sigma = \begin{pmatrix} \sigma_{xx} & \sigma_{xy} & \sigma_{xz} \\ \sigma_{yx} & \sigma_{yy} & \sigma_{yz} \\ \sigma_{zx} & \sigma_{zy} & \sigma_{zz} \end{pmatrix} \tag{10.9}$$

$$\sigma = - \begin{pmatrix} p & 0 & 0 \\ 0 & p & 0 \\ 0 & 0 & p \end{pmatrix} + \begin{pmatrix} \sigma_{xx} + p & \sigma_{xy} & \sigma_{xz} \\ \sigma_{yx} & \sigma_{yy} + p & \sigma_{yz} \\ \sigma_{zx} & \sigma_{zy} & \sigma_{zz} + p \end{pmatrix} \tag{10.10}$$

$$\sigma = -pI + \sigma_T \tag{10.11}$$

For the purpose of modeling, we often assume the fluids to be Newtonian, i.e., the stress is directly proportional to the rate of deformation. Therefore, the expression of stress for a Newtonian fluid can be given as

$$\sigma_{ij} = \mu \left(\frac{\partial u_i}{\partial x_j} + \frac{\partial u_j}{\partial x_i} \right) \tag{10.12}$$

In the above equation, μ is called the fluid viscosity. It defines the ease with which the fluid flows when subjected to body forces. The divergence of the stress term can now be stated as

$$\nabla \cdot \sigma = \mu \nabla \cdot \begin{pmatrix} \sigma_{xx} & \sigma_{xy} & \sigma_{xz} \\ \sigma_{yx} & \sigma_{yy} & \sigma_{yz} \\ \sigma_{zx} & \sigma_{zy} & \sigma_{zz} \end{pmatrix} \tag{10.13}$$

$$\nabla \cdot \sigma = \mu \begin{pmatrix} 2\frac{\partial u}{\partial x} & \frac{\partial u}{\partial y} + \frac{\partial v}{\partial x} & \frac{\partial u}{\partial z} + \frac{\partial w}{\partial x} \\ \frac{\partial u}{\partial y} + \frac{\partial v}{\partial x} & 2\frac{\partial v}{\partial y} & \frac{\partial v}{\partial z} + \frac{\partial w}{\partial y} \\ \frac{\partial u}{\partial z} + \frac{\partial w}{\partial x} & \frac{\partial v}{\partial z} + \frac{\partial w}{\partial y} & 2\frac{\partial w}{\partial z} \end{pmatrix} \tag{10.14}$$

The x component of the divergence term can be computed as follows:

$$(\nabla \cdot \sigma)_i = \mu \frac{\partial}{\partial x}\left(2\frac{\partial u}{\partial x}\right) + \frac{\partial}{\partial y}\left(\frac{\partial u}{\partial y}\frac{\partial v}{\partial x}\right) + \frac{\partial}{\partial z}\left(\frac{\partial u}{\partial z}\frac{\partial w}{\partial x}\right) \tag{10.15}$$

$$(\nabla \cdot \sigma)_i = \mu \frac{\partial^2 u}{\partial x^2} + \frac{\partial^2 u}{\partial y^2} + \frac{\partial^2 u}{\partial z^2} + \frac{\partial^2 u}{\partial x^2} + \frac{\partial^2 u}{\partial x \partial y} + \frac{\partial u^2}{\partial x \partial z} \tag{10.16}$$

$$(\nabla \cdot \sigma)_i = \mu \nabla^2 u + \mu \frac{\partial}{\partial x}\left(\frac{\partial u}{\partial x} + \frac{\partial v}{\partial y} + \frac{\partial w}{\partial z}\right) \tag{10.17}$$

$$(\nabla \cdot \sigma)_i = \mu \nabla^2 u + \mu \frac{\partial}{\partial x}(\nabla \cdot v) \tag{10.18}$$

$$(\nabla \cdot \sigma)_i = \mu \nabla^2 u + 0 \tag{10.19}$$

$$(\nabla \cdot \sigma)_i = \mu \nabla^2 u \tag{10.20}$$

The y and z terms can be computed in a similar manner. Consequently, for the case of incompressible Newtonian fluids, the flow equation can be given as

$$\rho \frac{D\mathbf{v}}{Dt} = -\nabla p + \mu \nabla^2 u + \mathbf{f} \tag{10.21}$$

We shall refer to this equation in the upcoming sections where we extend the derivation to Darcy flow and Saint-Venant equations. The derivations shall be followed by their implementation in Python programming language.

10.1 Numerical Modeling Using Finite Elements

The finite element models are a class of methods for solving partial differential equations. The method works by dividing a large system into smaller and simpler parts called finite elements. This is achieved by realizing a mesh over the computational domain where the differential equations need to be solved. The mesh helps define the points/nodes required for the discretization of the differential equation. The "finite" nature of the problem is a consequence of the finite number of nodes used in discretization. The important concepts required for their implementation are described next.

The basic approach to solving a partial differential equation using the finite element method can be given as follows:

(1) Express the given partial differential equation in its variational/weak form.
(2) Mesh the computational domain to discretize the weak form equations.
(3) Select the basis function whose derivative and values are simple to compute.
(4) Realize a linear algebraic system of equations to be solved using numerical solvers.

The product rule of divergence is given in Eq. 10.22. This is often useful in deriving the weak forms of partial differential equations (PDEs). In this equation, the divergence of the product of two functions is given as the sum of: (the product of the gradient of the function and the vector) and (product of the function and the divergence of the vector). The symbol $*$ denotes the product operation. This is of course true when the divergence of the vector is known.

$$\nabla \cdot (f * v) = (\nabla f) * v + f * (\nabla \cdot v) \tag{10.22}$$

We often deal with boundary value problems where the value of the unknown function is given at the boundary of the domain. The Divergence theorem comes in quite handy for assigning the values at the boundary:

$$\iiint_V (\nabla \cdot \mathbf{F}) \, dV = \oiint_{S(V)} (\mathbf{F} \cdot \hat{\mathbf{n}}) \, dS \tag{10.23}$$

A related formula often relevant for 1D analysis, called "*Integration by parts*" can be compactly stated as

$$\int u \, dv = uv - \int v \, du \tag{10.24}$$

where, $dv = v'(x) = \frac{dv}{dx}$ and $du = u'(x) = \frac{du}{dx}$.

10.2   Weak Form of the Steady State Darcy Flow Equation

For a single-phase flow problem Darcy's pressure law can be stated as in Eq. 10.25.

$$q = -\frac{k}{\mu L} \Delta p \tag{10.25}$$

Now, coming to our example of the derivation of the weak form of Darcy's pressure law. The law in Eq. 10.25 can be alternatively given in the form as in Eq. 10.26.

$$-\nabla \cdot \frac{K}{\mu}\nabla p = 0 \qquad \in \Omega \qquad (10.26)$$

Here, the first $(\nabla \cdot)$ symbol denotes the divergence operation, and the second (∇) before p denotes the gradient operation.

(1) The first step is to multiply Eq. 10.26 with a test function ψ.

$$\psi\left(-\nabla \cdot \frac{K}{\mu}\nabla p\right) = 0 \qquad \forall \psi \qquad (10.27)$$

One can easily realize the analogy between the LHS of Equation 10.27 and the second term in the RHS of Equation 10.22.

(2) The second step is to integrate the equation over the domain Ω.

$$-\int_\Omega \psi\left(\nabla \cdot \frac{K}{\mu}\nabla p\right) = 0 \qquad (10.28)$$

(3) The third step is to do integration by parts using Eq. 10.22.

$$-\left(\int_\Omega \nabla \cdot \left(\psi * \left(\frac{K}{\mu}\nabla p\right)\right) - \int_\Omega \nabla\psi * \left(\frac{K}{\mu}\nabla p\right)\right) = 0 \qquad (10.29)$$

Or,

$$\int_\Omega \nabla\psi * \left(\frac{K}{\mu}\nabla p\right) - \int_\Omega \nabla \cdot \left(\psi * \left(\frac{K}{\mu}\nabla p\right)\right) = 0 \qquad (10.30)$$

(4) The fourth step is to apply the divergence theorem on the second term in Eq. 10.30 using Eq. 10.23.

$$\int_\Omega \nabla\psi * \left(\frac{K}{\mu}\nabla p\right) - \oint_\Gamma \left(\psi * \left(\frac{K}{\mu}\nabla p\right)\right) \cdot \hat{n} = 0 \qquad (10.31)$$

(5) The fifth step is to express the Eq. 10.32 using inner product notation.

$$\left\langle \nabla\psi, \left(\frac{K}{\mu}\nabla p\right)\right\rangle - \left\langle \psi, \left(\frac{K}{\mu}\nabla p \cdot \hat{n}\right)\right\rangle = 0 \qquad (10.32)$$

Here, the first term is called a kernel and the second term is called the boundary condition.

10.3   Integration of Transient PDEs

The simplest of the PDE is a heat equation or a diffusion equation. A simple extension to the Poisson problem is the time-dependent heat equation or the time-dependent diffusion equation. A Poisson equation defines the stationary heat distribution in a body. This can be easily extended to a time-dependent problem. A straightforward approach to solving time-dependent PDEs is to first discretize the time derivative using finite-difference approximations which yields a sequence of stationary problems; then turning each stationary problem into a variational problem. Now, we illustrate a simple example:

$$\frac{\partial u}{\partial t} = \nabla^2 u + f \quad \text{in } \Omega \times (0, T] \tag{10.33}$$

$$u = u_D \quad \text{on } \partial\Omega \times (0, T] \tag{10.34}$$

$$u = u_0 \quad \text{at } t = 0 \tag{10.35}$$

Except, $u = u_0$, all others are a function of space and time $u = u(x, y, t)$. Even f is a function of space and time.

If n denotes a quantity at time t_n. For example, u_n means u at time level n. A finite-difference discretization would result into the following equations:

$$\left(\frac{\partial u}{\partial t}\right)^{n+1} = \nabla^2 u^{n+1} + f^{n+1} \tag{10.36}$$

A simple backward difference approximation leads to:

$$\frac{u^{n+1} - u^n}{\Delta t} = \nabla^2 u^{n+1} + f^{n+1} \tag{10.37}$$

This represents the time-discrete version of the above equation, also called as the *Backward-Euler* or *Implicit Euler* discretization. Therefore, assuming u^n is known, we can rewrite the equation as

$$u^{n+1} - \Delta t \nabla^2 u^{n+1} = u^n + \Delta t f^{n+1}, \qquad n = 0, 1, 2, \dots, n \tag{10.38}$$

Now, given u_0 we can solve for $u^0, u^1, u^2, \dots, u^n$. Collecting all the terms on the left-side gives us

$$u^{n+1} - \Delta t \nabla^2 u^{n+1} - u^n - \Delta t f^{n+1} = 0, \qquad n = 0, 1, 2, \dots, n \tag{10.39}$$

The above equation only has terms involving n and $n + 1$, therefore, if we know the value of u_n we can know u_{n+1}. With this, we have described the Backward-Euler time integration technique. There are advanced alternatives that offer advantages such as stability over time steps. Some examples of these methods are the Crank-Nicolson method, Adams-Bashforth, Adams-Moulton, Newmark Beta, and Runge-Kutta methods.

Chapter 11
Surface Flow Models

Surface flow models play a pivotal role in understanding and predicting hydrological processes, significantly influencing water resource management, flood risk assessment, and designing urban drainage systems. These models facilitate the understanding of spatial and temporal water flow patterns, addressing the complex interactions between the atmosphere, land surface, and subsurface. In recent years, the importance of surface flow models has grown, driven by increased urbanization, climate change impacts, and the need for sustainable water resource management. As we delve deeper into the world of surface flow models, kinematic wave approximations emerge as an essential tool for simplifying the modeling of flow processes. This approach allows for efficient and accurate simulations, particularly in open channel flow situations, where understanding the movement and storage of water is crucial for effective infrastructure planning and environmental conservation. By employing kinematic wave approximations, researchers and engineers can strike a balance between computational efficiency and model accuracy, ensuring the delivery of reliable solutions to pressing water management challenges.

The open channel flow can be given by the following equation:

$$\frac{\partial A}{\partial t} + \frac{\partial Q}{\partial x} = q \tag{11.1}$$

where, Q is the discharge flow rate, A is the cross-sectional area of the channel and q is the lateral inflow.

We can rewrite the first term in Eq. 11.1 as follows:

$$\frac{\partial Q}{\partial x} = \frac{\partial Q}{\partial A} \cdot \frac{\partial A}{\partial x} \tag{11.2}$$

Manning's friction law gives a relation between cross-sectional area and flow rate as follows:

© The Author(s), under exclusive license to Springer Nature Singapore Pte Ltd. 2024   183
A. Kumar and M. Saharia, *Python for Water and Environment*, Innovations in Sustainable Technologies and Computing, https://doi.org/10.1007/978-981-99-9408-3_11

$$Q = \frac{1}{n} A R^{2/3} S^{1/2} \tag{11.3}$$

where Q is the flow rate, A is the cross-sectional area, R hydraulic radius, and n, S are the Manning's roughness coefficient and channel slope, respectively. We can replace R with (A/P) to get the equation:

$$Q = \frac{1}{n} \frac{S^{1/2}}{P^{2/3}} A^{5/3} \tag{11.4}$$

where, P is the wetted parameter value.
   Then,

$$\frac{\partial Q}{\partial A} = \frac{S^{1/2}}{n P^{2/3}} \frac{\partial (A^{5/3})}{\partial A} \tag{11.5}$$

$$\frac{\partial Q}{\partial A} = \frac{S^{1/2}}{n P^{2/3}} \cdot \frac{5}{3} \cdot A^{\frac{5}{3}-1} \tag{11.6}$$

Substituting, Eq. 11.6 in Eq. 11.2, we get

$$\frac{\partial Q}{\partial x} = \frac{S^{1/2}}{n P^{2/3}} \cdot \frac{5}{3} \cdot A^{\frac{5}{3}-1} \frac{\partial A}{\partial x} \tag{11.7}$$

Substituting, Eq. 11.7 in Eq. 11.1 we get the general form of the kinematic wave approximation for an open channel.

$$\frac{\partial A}{\partial t} + \left(\frac{S^{1/2}}{n P^{2/3}} \right) \cdot \left(\frac{5}{3} \right) \cdot A^{\frac{5}{3}-1} \cdot \frac{\partial A}{\partial x} = q \tag{11.8}$$

A compact representation of the above Eq. 11.8 can be given as:

$$\frac{\partial A}{\partial t} + \alpha \beta A^{\beta-1} \frac{\partial A}{\partial x} = q \tag{11.9}$$

where $\alpha = \frac{S^{1/2}}{n P^{2/3}}$ and $\beta = 5/3$. One thing worth noticing in the above derivations is that we are solving for cross-sectional area, A. For computing the discharge, Q, we can use the following formula.

$$Q = \alpha A^{\beta} \tag{11.10}$$

The parameters n, S, P change for channels for different geometry. Next, we describe a finite-difference-based code implementation for solving Eq. 11.9 for a rectangular channel. Using the Taylor series expansion, we can write an approximation for $\frac{\partial A}{\partial t}$ and $\frac{\partial A}{\partial x}$ as follows:

$$\frac{\partial A}{\partial t} \approx \frac{A_{(i,j+1)} - A_{(i,j)}}{\Delta t} \tag{11.11}$$

$$\frac{\partial A}{\partial x} \approx \frac{A_{(i,j+1)} - A_{(i-1,j+1)}}{\Delta x} \tag{11.12}$$

Substituting, these into Eq. 11.9, we get the following:

$$\frac{A_{(i,j+1)} - A_{(i,j)}}{\Delta t} + \alpha\beta A_{mean}\frac{A_{(i,j+1)} - A_{(i-1,j+1)}}{\Delta x} = q \tag{11.13}$$

where, A_{mean} is given by:

$$A_{mean} = \frac{(A_{(i,j)} + A_{(i-1,j+1)})}{2}. \tag{11.14}$$

After some algebraic manipulations and making $A_{(i,j+1)}$ the subject of the formula, we get the following recurrence relation:

$$A_{(i,j+1)} = \frac{\Delta x q + \frac{\Delta x}{\Delta t}(A_{mean}) + \alpha\beta(A_{mean})A_{(i-1,j+1)}}{\frac{\Delta x}{\Delta t} + \alpha\beta(A_{mean})} \tag{11.15}$$

11.1 Rectangular Channel

Solving the kinematic wave approximation for rectangular channels requires understanding the channel geometry and flow characteristics. The channel's width and depth are constant, simplifying the cross-sectional area and wetted perimeter calculations. The approximation relies on continuity and momentum equations, incorporating assumptions such as steady, uniform flow, negligible pressure, and friction effects. By integrating these equations, one can derive a simplified form of the Saint-Venant equation that governs the flow. Consequently, the kinematic wave approximation allows for practical flow depth, velocity, and discharge calculations, making it an invaluable tool for analyzing rectangular channel hydraulics.

11.1.1 Python Code

The given code block is a simulation program for the kinematic wave equation in a rectangular channel. It numerically solves the kinematic wave equation using finite differences to simulate the flow behavior over time.

From lines 1–28, we begin by setting up the simulation parameters, such as the spatial and temporal grid dimensions, channel characteristics, and the Courant-Friedrichs-Lewy (CFL) condition. The CFL condition ensures the stability of the

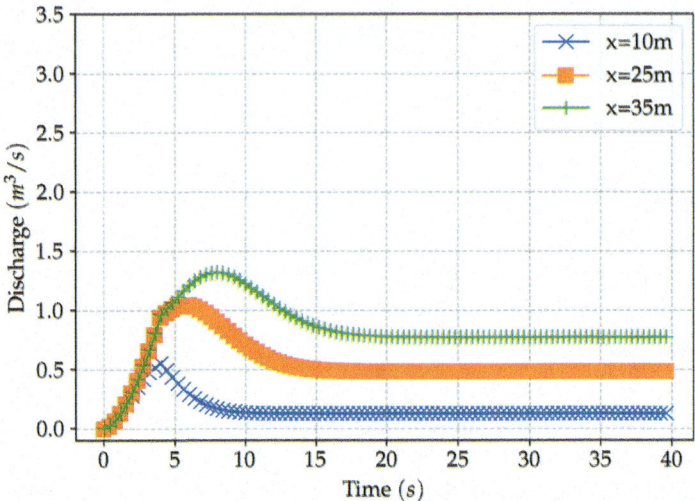

Fig. 11.1 Discharge as a function of time for the rectangular channel case. Moving away from the starting of the channel the amplitude of the wave increases

numerical scheme by limiting the time step size based on the grid spacing and the flow velocity.

Next, in lines 35–41, the code defines the model parameters like the roughness coefficient, channel slope, and width. It also specifies the rainfall or lateral inflow rate and the time at which the inflow occurs.

In lines 48–52, the solution variables, A (cross-sectional area) and t_{vals} (time values) are initialized. The initial condition for A is set to a constant value across the channel.

In line 55, the code then enters a time loop and calculates the cross-sectional area at each spatial point and time step, based on the kinematic wave equation. It considers the inflow conditions and the flow characteristics in the channel. The computed areas are used to calculate the discharge (Q) using the specified power-law relationship in line 76.

Finally, from lines 78–91, the code plots (Fig. 11.1) the discharge at specific spatial points over time and saves the plot as an image file.

In line 93, if the CFL condition is violated and the time step size is more than the time step calculated from the CFL condition, the code exits with an error message.

This way the code simulates the kinematic wave equation in a rectangular channel, computes the cross-sectional area and discharge, and visualizes the discharge over time at different spatial locations.

```python
"""
Program to simulate the kinematic wave equation
for a rectangular channel
"""

import numpy as np
import matplotlib.pyplot as plt
plt.rc('text', usetex=True)

"""
Simulation parameters
"""
x_start = 0  # (m)
x_stop = 100  # (m)
NX = 50  # number of spatial grid point
x, dx = np.linspace(x_start, x_stop, NX, retstep=True)

t_start = 0  # (s)
t_stop = 40  # (s)
NT = 100  # number of temporal grid points
dt = (t_stop - t_start) / NT

"""
Courant_Friedrichs Lewy (CFL) condition
"""
CFL = 0.5
v = 1.0
dt_cfl = CFL * dx / np.abs(v)

if dt <= dt_cfl:
    print("dt < dt_CFL")
    print(dt, "<", dt_cfl)
    print("Algorithm is stable - proceed")

    # Setting model parameters
    sim_name = "Rect"
    n = 0.025  # roughness coefficient
    S = 0.01  # slope of channel
    W = 9.00  # width of channel
    alpha = (1.49/n) * S**(1/2) * W**(-2/3)
```

```
41      beta = 5 / 3
42
43      # Rainfall or lateral inflow
44      inflow = 0.1   # (m^3 / s)
45      time_to_inflow = 10   # (s)
46
47      # Solution variables
48      A = np.zeros([NX, NT])   # time along the columns
49      t_vals = np.zeros([NT, ])   # storing in actual units
50
51      # Initial condition
52      A[:, 0] = 0.005
53
54      # Time loop
55      for t in range(0, NT - 1):
56          # Storing the time values in seconds
57          t_vals[t] = t * dt
58
59          # Lower inflow to occur after some time
60          if t < time_to_inflow:
61              q = inflow
62          else:
63              q = inflow / 3
64
65          # Space loop
66          for x in range(1, NX):
67              A_mean = 0.5 * (A[x, t] + A[x - 1, t + 1])
68              A[x, t + 1] = np.divide(dx*q + (dx/dt)
69                                      * A_mean +
70                                      alpha*beta*A_mean*
71                                      A[x - 1, t + 1],
72                                      (dx/dt) +
73                                      alpha*beta*A_mean)
74
75      # Computing discharge based on area
```

```
76          Q = alpha * A**beta
77
78          # Plot
79          t_vals[-1] = t_vals[-2] + dt
80          plt.plot(t_vals, Q[int(10), :], "-x",
81                      label="x=" + str(int(10)) + "m")
82          plt.plot(t_vals, Q[int(NX / 2), :], "-s",
83                      label="x=" + str(int(NX / 2)) + "m")
84          plt.plot(t_vals, Q[int(NX / 2) + 10, :], "-+",
85                      label="x=" + str(int(NX / 2) + 10) + "m")
86          plt.xlabel("Time $(s)$")
87          plt.ylabel("Discharge $(m^3/s)$")
88          plt.legend()
89          plt.grid(ls="--")
90          plt.tight_layout()
91          plt.savefig("./result/" + sim_name + "_.png", dpi=300)
92
93      else:
94          print("dt > dt_CFL")
95          print(dt, ">", dt_cfl)
96          print("Unstable - reduce dt!")
97          print("Program exiting!")
98          exit()
99
100     print("Done!")
```

11.2   Triangular Channel

Triangular channels, characterized by their V-shaped cross-section, present a unique geometry that influences the behavior of flow within the channel. When applying the kinematic wave approximation to triangular channels, the channel's side slope and flow characteristics must be taken into account. The method simplifies the governing equations by assuming uniform flow, steady-state conditions, and negligible pressure and friction effects. Solving these simplified equations yields the flow depth, velocity, and discharge in the triangular channel. The kinematic wave approximation proves advantageous in these settings, offering rapid and accurate insights into the hydraulic behavior of triangular channels.

11.2.1   Python Code

In this section, we describe the code to solve the kinematic wave equation for a triangular channel. It is similar in all respects to the rectangular channel case except in the definition of the model parameters. The code aims to compute the discharge at different positions along the channel over a specified time period.

In lines 13–28, we define simulation parameters such as spatial and temporal domain, number of grid points, and also define time step. The Courant-Friedrichs-Lewy (CFL) condition is calculated to ensure the stability of the algorithm.

Next, in lines 35–43, the model parameters for the triangular channel are set, including the roughness coefficient, slope of the channel, bottom width, and side slope. The alpha coefficient is computed based on these parameters.

In lines 45–51, the code defines the rainfall or lateral inflow values and the time at which the inflow occurs. Solution variables for the channel area and time values are initialized.

The time loop starts at line 56, which iterates over the specified number of time steps. Within each time step, the space loop iterates over the spatial grid points. The kinematic wave equation is solved using the numerical scheme, and the updated channel area is stored.

After the time loop, the discharge is computed at line 78, based on the area using the alpha and beta coefficients.

Finally, in lines 80–93, the code plots (Fig. 11.2) the discharge at specific positions along the channel over time. The discharge values at different grid points are plotted against the time values. The resulting plot shows the variation in discharge over time for different positions along the channel.

The code includes checks to ensure stability by comparing the time step with the CFL condition in line 30. In line 95, if the time step is larger than the CFL condition, the program exits with a message indicating the need to reduce the time step for stability.

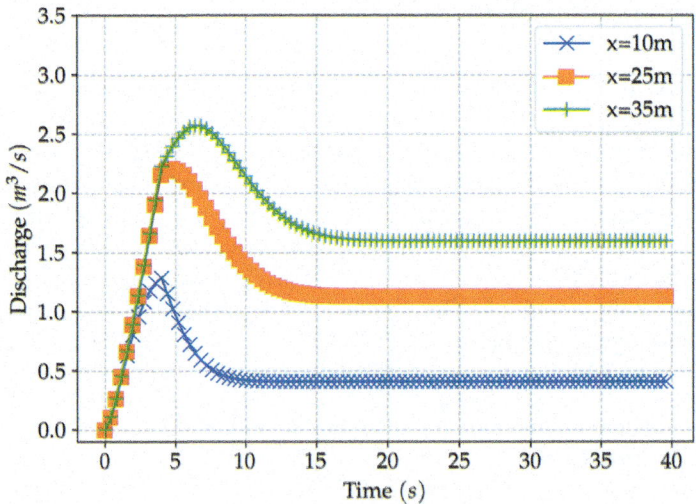

Fig. 11.2 Discharge as a function of time for the triangular channel case. Moving away from the starting of the channel the amplitude of the wave increases

Overall, the code provides a simulation of the kinematic wave equation for a triangular channel and visualizes the discharge variation over time at different positions along the channel.

```
1    """
2    Program to simulate the kinematic wave equation
3    for a triangular channel
4    """
5
6    import numpy as np
7    import matplotlib.pyplot as plt
8    plt.rc('text', usetex=True)
9
10   """
11   Simulation parameters
12   """
13   x_start = 0   # (m)
14   x_stop = 100   # (m)
15   NX = 50   # number of spatial grid point
16   x, dx = np.linspace(x_start, x_stop, NX, retstep=True)
17
18   t_start = 0   # (s)
19   t_stop = 40   # (s)
20   NT = 100   # number of temporal grid points
21   dt = (t_stop - t_start) / NT
22
23   """
24   Courant_Friedrichs Lewy (CFL) condition
25   """
26   CFL = 0.5
27   v = 1.0
28   dt_cfl = CFL * dx / np.abs(v)
29
30   if dt <= dt_cfl:
31       print("dt < dt_CFL")
32       print(dt, "<", dt_cfl)
33       print("Algorithm is stable - proceed")
34
35       # Setting model parameters
36       sim_name = "Tri"
37       n = 0.025   # roughness coefficient
38       S = 0.01   # slope of channel
39       b = 2.0   # channel bottom width
40       z = 1   # 45 degrees side slope
41       Z = b/(2*z)
42       alpha = (0.94/n) * S**(1/2) * (Z/(1+Z**2))**(1/3)
```

```
43    beta = 4 / 3
44
45    # Rainfall or lateral inflow
46    inflow = 0.1  # (m^3 / s)
47    time_to_inflow = 10  # (s)
48
49    # Solution variables
50    A = np.zeros([NX, NT])  # time along the columns
51    t_vals = np.zeros([NT, ])  # storing in actual units
52
53    # Initial condition
54    A[:, 0] = 0.005
55
56    # Time loop
57    for t in range(0, NT - 1):
58        # Storing the time values in seconds
59        t_vals[t] = t * dt
60
61        # Lower inflow to occur after some time
62        if t < time_to_inflow:
63            q = inflow
64        else:
65            q = inflow / 3
66
67        # Space loop
68        for x in range(1, NX):
69            A_mean = 0.5 * (A[x, t] + A[x - 1, t + 1])
70            A[x, t + 1] = np.divide(dx*q + (dx/dt)
71                                    * A_mean +
72                                    alpha*beta*A_mean*
73                                    A[x - 1, t + 1],
74                                    (dx/dt) +
75                                    alpha*beta*A_mean)
76
77    # Computing discharge based on area
78    Q = alpha * A**beta
79
80    # Plot
81    t_vals[-1] = t_vals[-2] + dt
82    plt.plot(t_vals, Q[int(10), :], "-x",
83             label="x=" + str(int(10)) + "m")
```

```
84      plt.plot(t_vals, Q[int(NX / 2), :], "-s",
85              label="x=" + str(int(NX / 2)) + "m")
86      plt.plot(t_vals, Q[int(NX / 2) + 10, :], "-+",
87              label="x=" + str(int(NX / 2) + 10) + "m")
88      plt.xlabel("Time $(s)$")
89      plt.ylabel("Discharge $(m^3/s)$")
90      plt.legend()
91      plt.grid(ls="--")
92      plt.tight_layout()
93      plt.savefig("./result/" + sim_name + "_.png", dpi=300)
94
95  else:
96      print("dt > dt_CFL")
97      print(dt, ">", dt_cfl)
98      print("Unstable - reduce dt!")
99      print("Program exiting!")
100     exit()
101
102 print("Done!")
```

11.3 Circular Channel

Circular channels, with their curved cross-sections, necessitate special attention when applying the kinematic wave approximation. The channel's diameter, flow conditions, and hydraulic radius must be considered in order to accurately represent the channel's geometry. By assuming uniform flow, steady-state conditions, and disregarding pressure and friction effects, the governing equations can be simplified for circular channels. Upon solving these equations, one can estimate the flow depth, velocity, and discharge within the channel. The kinematic wave approximation, despite the unique challenges posed by circular channels, continues to be an effective method for rapidly approximating flow characteristics in these systems.

11.3.1 Python Code

This program is similar to the previous rectangular and triangular channel cases. In lines 13–28, it starts by setting up the simulation parameters such as the spatial and temporal domain, number of grid points, and the CFL condition. The CFL condition is used to ensure the stability of the numerical scheme.

Next, in lines 36–41, the model parameters such as roughness coefficient, slope of the channel, and diameter of the channel are defined. These parameters are used to calculate the alpha coefficient, a function of the roughness, slope, and diameter. Additionally, the inflow rate and time at which the inflow starts are specified.

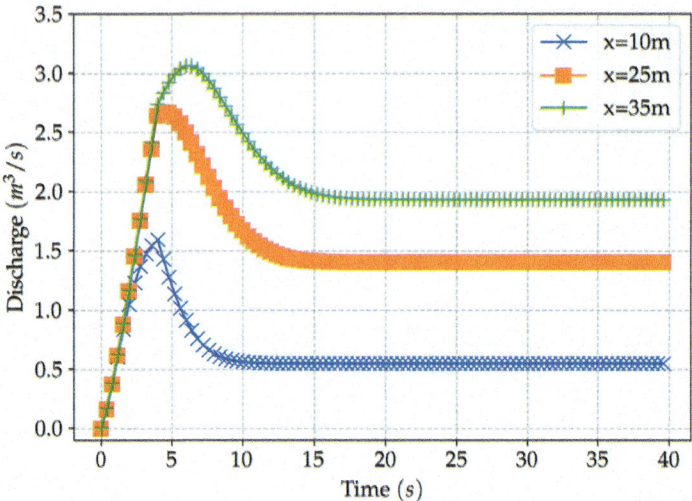

Fig. 11.3 Discharge as a function of time for the circular channel case. Moving away from the starting of the channel the amplitude of the wave increases

The solution variables, including the cross-sectional area and time values, are initialized in lines 48–49. The initial condition for the cross-sectional area is set to a constant value. The time loop starts at line 55 which iterates over the specified number of time steps, and within each time step, the space loop, starting at line 66, updates the cross-sectional area based on the kinematic wave equation. The computation involves calculating the mean area, inflow rate, and the new area based on the numerical scheme.

In line 76, the discharge is computed based on the updated cross-sectional area using the alpha and beta coefficients. Finally, the code plots (Fig. 11.3) the discharge at different spatial locations over time and saves the figure in lines 78–91.

The code includes checks for stability using the CFL condition in line 30. If the condition is satisfied, the simulation proceeds; otherwise, a warning is displayed, and the program exits using line 93. The program also provides visual outputs to analyze the variation of discharge at different spatial locations over time.

The code solves the kinematic wave equation for a circular channel, allowing for the study of flow dynamics and discharge variations under specified conditions.

```python
"""
Program to simulate the kinematic wave equation
for a circular channel
"""

import numpy as np
import matplotlib.pyplot as plt
plt.rc('text', usetex=True)

"""
Simulation parameters
"""
x_start = 0  # (m)
x_stop = 100  # (m)
NX = 50  # number of spatial grid point
x, dx = np.linspace(x_start, x_stop, NX, retstep=True)

t_start = 0  # (s)
t_stop = 40  # (s)
NT = 100  # number of temporal grid points
dt = (t_stop - t_start) / NT

"""
Courant_Friedrichs Lewy (CFL) condition
"""
CFL = 0.5
v = 1.0
dt_cfl = CFL * dx / np.abs(v)

if dt <= dt_cfl:
    print("dt < dt_CFL")
    print(dt, "<", dt_cfl)
    print("Algorithm is stable - proceed")

    # Setting model parameters
    sim_name = "Circ"
    n = 0.025  # roughness coefficient
    S = 0.01  # slope of channel
    D = 2.00  # diameter of channel
    alpha = (0.804/n) * S**(1/2) * D**(1/6)
```

```
41    beta = 5 / 4
42
43    # Rainfall or lateral inflow
44    inflow = 0.1   # (m^3 / s)
45    time_to_inflow = 10   # (s)
46
47    # Solution variables
48    A = np.zeros([NX, NT])   # time along the columns
49    t_vals = np.zeros([NT, ])   # storing in actual units
50
51    # Initial condition
52    A[:, 0] = 0.005
53
54    # Time loop
55    for t in range(0, NT - 1):
56        # Storing the time values in seconds
57        t_vals[t] = t * dt
58
59        # Lower inflow to occur after some time
60        if t < time_to_inflow:
61            q = inflow
62        else:
63            q = inflow / 3
64
65        # Space loop
66        for x in range(1, NX):
67            A_mean = 0.5 * (A[x, t] + A[x - 1, t + 1])
68            A[x, t + 1] = np.divide(dx*q + (dx/dt)
69                                     * A_mean +
70                                     alpha*beta*A_mean*
71                                     A[x - 1, t + 1],
72                                     (dx/dt) +
73                                     alpha*beta*A_mean)
74
75    # Computing discharge based on area
76    Q = alpha * A**beta
77
78    # Plot
79    t_vals[-1] = t_vals[-2] + dt
80    plt.plot(t_vals, Q[int(10), :], "-x",
81             label="x=" + str(int(10)) + "m")
```

```
82      plt.plot(t_vals, Q[int(NX / 2), :], "-s",
83              label="x=" + str(int(NX / 2)) + "m")
84      plt.plot(t_vals, Q[int(NX / 2) + 10, :], "-+",
85              label="x=" + str(int(NX / 2) + 10) + "m")
86      plt.xlabel("Time $(s)$")
87      plt.ylabel("Discharge $(m^3/s)$")
88      plt.legend()
89      plt.grid(ls="--")
90      plt.tight_layout()
91      plt.savefig("./result/" + sim_name + "_.png", dpi=300)
92
93  else:
94      print("dt > dt_CFL")
95      print(dt, ">", dt_cfl)
96      print("Unstable - reduce dt!")
97      print("Program exiting!")
98      exit()
99
100 print("Done!")
```

11.3.2 Discussion of Results

We now compare the results of 1D kinematic wave approximation for the three-channel cases, viz, the rectangular, triangular, and circular channels. All the simulation parameters were the same for the three cases except the expressions for the α, β. Upon comparing Figs. 11.1, 11.2, and 11.3 one can see that the height of the wave is the most significant feature. For the rectangular case, the height is lowest owing to the width of the channel that furnishes to accommodate low height by providing space for the water to stay near the surface. Interestingly, in Fig. 11.2 the waves tend to show a peaked pattern, owing to the narrow base of the triangular channel. Similarly, in Fig. 11.3, the wave height still peaks more, owing to the overall decrease in the flow area. The three plots show the wave height as a function of time at three locations in the 1D domain. These types of simulations can help study various scenarios where the geometry of the flow channel varies.

11.4 2D Shallow Water Equations

The 2D shallow water equations (SWE) are a set of partial differential equations that describe the behavior of fluid flow in a shallow water environment, where the depth of the fluid is lesser compared to its horizontal extent. In the context of surface flow models, these equations are instrumental in simulating and predicting various hydrodynamic processes, such as overland flow, river flow, and coastal inundation, where the fluid depth variations are significant compared to the flow velocities.

The 2D SWE consists of three primary equations representing the conservation of mass (continuity equation) and momentum in horizontal directions (x and y). These equations account for fluid depth, flow velocity, gravitational force, and bed elevation, comprehensively representing fluid dynamics in two horizontal dimensions. Due to their ability to capture complex flow phenomena, 2D SWE is widely employed in surface flow models, offering a detailed understanding of flow patterns, velocities, and depths in various hydrological systems.

The Saint-Venant equations, which form the basis of the 1D shallow water equations, are closely related to the 2D SWE. Both sets of equations are derived from the same fundamental principles of mass and momentum conservation, with the primary difference being the dimensionality of the flow representation. While the Saint-Venant equations are confined to one spatial dimension and are ideal for simulating flow in channels with relatively simple geometries, the 2D SWE provides a more comprehensive representation of flow patterns in situations with complex spatial variations, such as floodplains, estuaries, and urban areas.

Applications of the 2D shallow water equations and the Saint-Venant equations are widespread in studying surface flow models. They are used for predicting and managing flood risks, designing hydraulic structures, assessing the impact of land-use changes, and evaluating the consequences of extreme hydrological events. Moreover, these equations have been incorporated into coupled surface-subsurface models, enhancing the understanding the interactions between groundwater and surface water systems.

Thus, the 2D shallow water and Saint-Venant equations play a pivotal role in studying surface flow models. Their ability to capture the essential hydrodynamic processes in various hydrological systems makes them indispensable tools for researchers, engineers, and water resource managers seeking to understand and address the challenges associated with water flow, flood risk, and environmental conservation.

11.4.1   Governing Equations

The governing equations for the 2D shallow water equation is given as:

$$\frac{\partial h}{\partial t} + \frac{\partial hu}{\partial x} + \frac{\partial hv}{\partial y} = 0 \quad \text{in } \Omega \times [0, T] \tag{11.16}$$

$$\frac{\partial hu}{\partial t} + \frac{\partial hu^2 + 0.5gh^2}{\partial x} + \frac{\partial huv}{\partial y} = 0 \quad \text{in } \Omega \times [0, T] \tag{11.17}$$

$$\frac{\partial hv}{\partial t} + \frac{\partial huv}{\partial x} + \frac{\partial hv^2 + 0.5gh^2}{\partial y} = 0 \quad \text{in } \Omega \times [0, T] \tag{11.18}$$

where, Ω, T denote the computational domain and T the total time for which the simulation runs. h, u, v are the height, x-velocity, and y-velocity of the water. g

denotes gravity. We apply the reflecting boundary conditions on all four sides of the rectangular domain.

11.4.2   Python Code

This code implements a shallow water simulation using the Finite Volume Method (FVM). The simulation aims to model water flow behavior in a two-dimensional domain.

From lines 13–37, the code begins with the file I/O setup, including defining the simulation name and result path. It also describes the domain size, grid resolution, and the parameters for the simulation, such as gravity (g) and the Courant-Friedrichs-Lewy (CFL) number.

Next, from lines 51–77, the code initializes the variables and arrays needed for the simulation. It sets up the grid by creating mesh grid points in the x and y directions, representing the domain. The initial values for the water height (eta) are defined using a function "eta_initial()", which can be modified according to the desired initial conditions.

In lines 82–129, the simulation proceeds with a time loop, iterating until a specified end time (t_stop) is reached. Inside the loop, the simulation time (t_curr) and time step (dt) are updated. The alpha values are calculated based on the water height and velocity.

From lines 141–175, the code computes the fluxes using the Lax-Friedrichs scheme, which is a numerical scheme for approximating the fluxes in shallow water equations. The fluxes are computed in both the x and y directions separately.

Next, in lines 177–180, the "Solution" variable is updated using the computed fluxes, accounting for the time step and grid spacing. In lines 182–197, boundary conditions are applied to enforce no-slip conditions on the water height and velocities at the boundaries.

In lines 200–201, the code calculates the velocities (u and v) by dividing the momentum components by the water height. It also performs necessary array assignments for the next time step.

During the simulation, the water height at each time step is stored in the "etas" list. Additionally, the code generates visualization outputs (.vtk) using the PyVista library. It creates structured grids and saves them in VTK format for later visualization and analysis.

Therefore, the code performs a shallow water simulation using the Finite Volume Method. It iterates over time steps, updates the solution variables, and computes the fluxes based on the shallow water equations. Visualization outputs are generated to visualize the water height and velocities during the simulation.

```
1    """
2    Shallow water simulation using the
3    Finite Volume Method (FVM) in
4    rectangular domain
5    """
6
7    import numpy as np
8    import pyvista as pv
9
10   """
11   File I/O
12   """
13   simulationName = "swe_FVM"
14   resultPath = "./result/" + simulationName + "/"
15
16
17   """
18   Computational domain creation
19   """
20   xlen = 2
21   ylen = 7
22   divs = 60
23   dx = xlen / divs
24   dy = ylen / divs
25
26   x = np.arange(-1 - dx, 1 + dx, dx)
27   y = np.arange(-1 - dy, 1 + dy, dy)
28   xx, yy = np.meshgrid(x, y)
29   print("Min {}, {}".format(np.min(xx), np.min(yy)))
30   print("Max {}, {}".format(np.max(xx), np.max(yy)))
31
32
33   """
34   Model parameters
35   """
36   g = 9.8
37   cfl = 0.5
38
39
40   # Source
41   def eta_initial(y, x):
42       """Intial eta values function."""
43       Amp = 8.0
```

```
44        x, y = -0.1, -0.9
45          = 0.3
46        return Amp * np.exp(
47            -1 * (0.02*(x - x) ** 2 +
48                    (y - y) ** 2) / ( ** 2))
49
50
51    # Base depth + Gaussian amplitude
52    eta = np.ones_like(xx)
53    eta = eta + eta_initial(yy, xx)
54
55    # Solution variable initializing as [mesh_size x 3] shape
56    Solution = np.zeros([xx.shape[0], xx.shape[1], 3])
57    Solution[:, :, 0] = eta   # Zero index corresponds to etas
58    u = np.zeros_like(xx)
59    v = np.copy(u)
60
61    # Shifting the indices by 1 for easing the BC application
62    p_p1_x = np.roll(np.arange(len(x)), 1)
63    p_p1_y = np.roll(np.arange(len(y)), 1)
64    p_m1_x = np.roll(np.arange(len(x)), -1)
65    p_m1_y = np.roll(np.arange(len(y)), -1)
66
67    # Running parameters
68    t_curr = 0.0
69    dt = 0.0
70    t_stop = 2.0
71
72    # Solution variables
73    etas = list()
74    times = list()
75    times.append(t_curr)
76    Solution_old = Solution  # old solution variable
77    counter = 0
78
79    """
80    Time-marching
81    """
82    while t_curr < t_stop:
83        # Save files
84        hsol = Solution[:, :, 0]
```

```
85      grid = pv.StructuredGrid(xx * xlen, yy * ylen, hsol)
86      grid.point_data["height"] = hsol.flatten(order="F")
87
88      top = grid.points.copy()
89      bottom = grid.points.copy()
90      bottom[:, -1] = -5.0  # Bottom plane
91      vol = pv.StructuredGrid()
92      vol.points = np.vstack([top, bottom])
93      vol.dimensions = [*grid.dimensions[0:2], 2]
94
95      z_above = hsol.flatten(order="F")
96      zlevels = vol.points[:, 2]
97      zlevels[zlevels < np.min(z_above)] = -0.5
98      vol.point_data["height"] = zlevels
99
100     write_format = resultPath + \
101                    "height_" + \
102                    "{:04d}.vtk".format(counter)
103     vol.save(write_format)
104
105
106     # Initialising variables
107     u_old = u  # old x-velocity
108     v_old = v  # old y-velocity
109
110     t_curr = t_curr + dt  # incrementing time
111
112     times.append(t_curr)
113
114     # calculate alpha = abs(u) + sqrt(gh) used for flux
115     alpha_u = 0.5 * np.abs(u_old +
116                            u_old[:, p_m1_x]) + \
117             np.sqrt(
118         g * 0.5 * (Solution_old[:, :, 0] +
119                    Solution_old[:, p_m1_x, 0]))
120     alpha_v = 0.5 * np.abs(v_old +
121                            v_old[p_m1_y, :]) + \
122             np.sqrt(
123         g * 0.5 * (Solution_old[:, :, 0] +
124                    Solution_old[p_m1_y, :, 0]))
125
126     # compute maximum alpha
```

```
127      alpha_max = np.linalg.norm(
128          x=np.hstack([np.ravel(alpha_u), np.ravel(alpha_v)]),
129          ord=np.inf)
130
131      # computing dt on the fly for stability
132      dt = np.min([cfl * (dx / alpha_max),
133                   cfl * (dy / alpha_max)])
134
135      # pre-computing some terms
136      huv = Solution_old[:, :, 1] * Solution_old[:, :, 2] / \
137          Solution_old[:, :, 0]
138      gh_sqr = 0.5 * g * Solution_old[:, :, 0] ** 2
139
140      # compute (hu, hu ** 2 + gh ** 2 / 2, huv)
141      LFFlux_u = np.stack([Solution_old[:, :, 1],
142                           Solution_old[:, :, 1] ** 2 /
143                           Solution_old[:, :, 0] + gh_sqr,
144                           huv],
145                          2)
146
147      # compute (hv, huv, hv ** 2 + gh ** 2 / 2)
148      LFFlux_v = np.stack([Solution_old[:, :, 2],
149                           huv,
150                           Solution_old[:, :, 2] ** 2. /
151                           Solution_old[:, :, 0] + gh_sqr],
152                          2)
153
154      # compute fluxes in x, y direction
155      flux_x = np.zeros_like(LFFlux_u)
156      flux_y = np.zeros_like(LFFlux_u)
157      for ii in range(flux_x.shape[2]):
158          temp_LFFluxu = LFFlux_u[:, p_m1_x, :]
159          temp_Uold = Solution_old[:, p_m1_x, :] - \
160                      Solution_old
161          flux_x[:, :, ii] = 0.5 * (
162                  LFFlux_u[:, :, ii] +
163                  temp_LFFluxu[:, :, ii]) - \
164                          0.5 * np.multiply(
165              temp_Uold[:, :, ii], alpha_u)
166
167      for ii in range(flux_y.shape[2]):
168          temp_LFFluxv = LFFlux_v[p_m1_y, :, :]
169          temp_Uold = Solution_old[p_m1_y, :, :] - \
```

```
170                    Solution_old
171        flux_y[:, :, ii] = 0.5 * (
172              LFFlux_v[:, :, ii] +
173              temp_LFFluxv[:, :, ii]) - \
174                    0.5 * np.multiply(
175        temp_Uold[:, :, ii], alpha_v)
176
177    # assiging to solution variable
178    Solution = Solution_old - \
179            (dt / dx) * (flux_x - flux_x[:, p_p1_x, :]) -\
180            (dt / dy) * (flux_y - flux_y[p_p1_y, :, :])
181
182    # imposing no-slip boundary conditions on eta, hu and hv,
183    # respectively
184    Solution[0:, -1, 0] = Solution[0:, -2, 0]
185    Solution[0:, 0, 0] = Solution[0:, 1, 0]
186    Solution[-1, 0:, 0] = Solution[-2, 0:, 0]
187    Solution[0, 0:, 0] = Solution[1, 0:, 0]
188
189    Solution[0:, -1, 1] = -Solution[0:, -2, 1]
190    Solution[0:, 0, 1] = -Solution[0:, 1, 1]
191    Solution[-1, 0:, 1] = Solution[-2, 0:, 1]
192    Solution[0, 0:, 1] = Solution[1, 0:, 1]
193
194    Solution[0:, -1, 2] = Solution[0:, -2, 2]
195    Solution[0:, 0, 2] = Solution[0:, 1, 2]
196    Solution[-1, 0:, 2] = -Solution[-2, 0:, 2]
197    Solution[0, 0:, 2] = -Solution[1, 0:, 2]
198
199    # Retrieving u and v
200    u = np.divide(Solution[:, :, 1], Solution[:, :, 0])
201    v = np.divide(Solution[:, :, 2], Solution[:, :, 0])
202
203    print("Done timestep {:.04f}".format(t_curr))
204    etas.append(Solution[:, :, 0])
205    Solution_old = Solution  # Assigning new solution to old
206    counter = counter + 1
207
208 print("Done!")
```

11.4.3 Discussion of the Results

The dam break simulation is an interesting application of 2D shallow water equation (SWE). The solutions of the SWE can provide insights into the occurrence of the sudden and catastrophic nature of the dam. The real-world dam break situation involves reflections, water wave propagation, and interaction with the obstacles—the SWE can handle both the flow velocity and depth making it possible to model such scenarios.

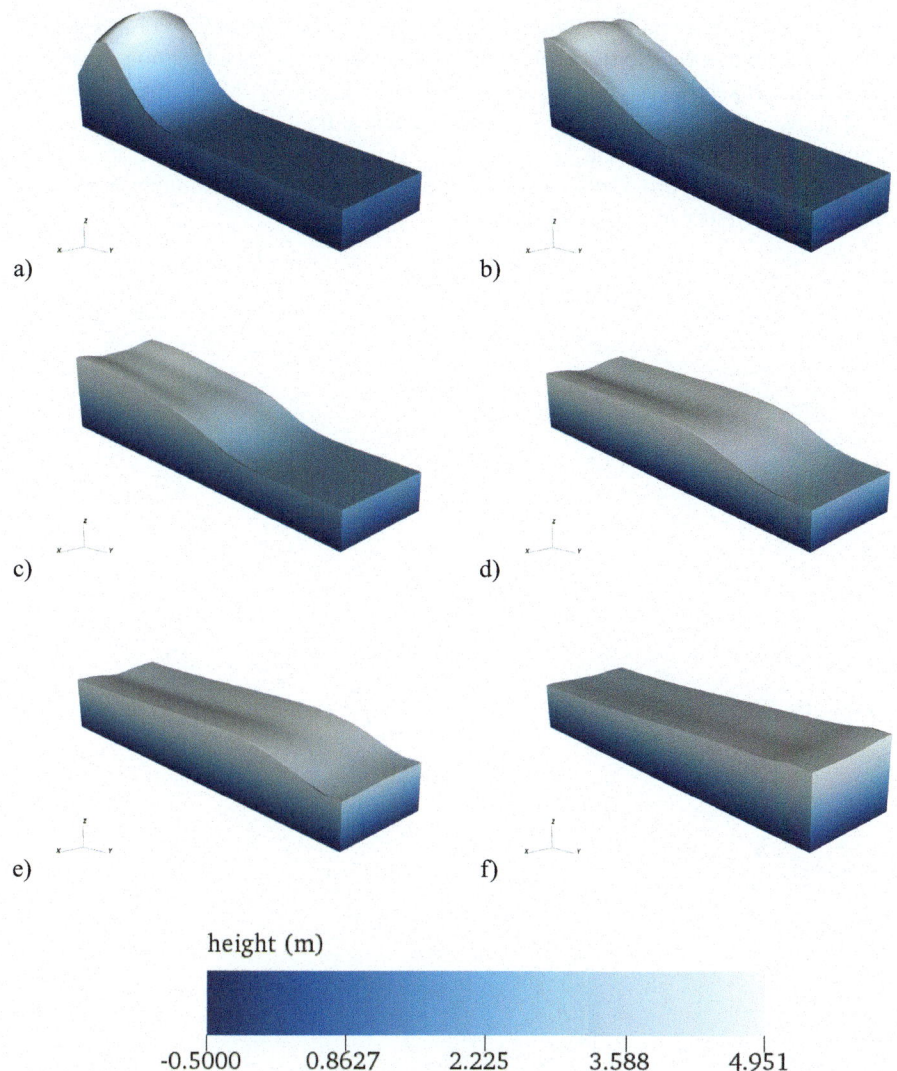

a)

b)

c)

d)

e)

f)

height (m)

-0.5000   0.8627   2.225   3.588   4.951

Fig. 11.4 Dam-break simulation. Solution of the shallow water equations is shown for the height. On a 2D domain at a point toward the far end of the computational domain, the height of the water is initially kept high while the other sides are kept lower. The hyperbolic PDE is solved until t_{final}. From **a** to **f** Instances of the height field at multiple times starting from t_0

For this simulation, the computational domain was created within the program itself. The relevant lines in the code are given in 26–28. The lines define a rectangular domain. The number of divisions in each axis is taken care of by the parameters "dx", and "dy" as shown in lines 23–24. The 2D shallow water equations were solved for this computational domain and the results of which are shown in Fig. 11.4. From (a) to (f) in the figures we see the evolution of the water height as a function of time.

Subfigure (a) shows the initial state of the water height. With this initial condition the shallow water equations are solved to reach a final time, t_{final}. Each of the subfigures clearly depicts the growing and diminishing waves. Hence, simulation can help simulate hydrograph sensor data when probed with a suitable number of sensors at different parts of the domain.

Chapter 12
Subsurface Flow Models

12.1 Seepage Flow Model

Seepage flow models form an integral part of subsurface hydrological studies, focusing on the simulation and quantification of fluid movement through porous media such as soil, rock, and sediment. These models are critical for understanding groundwater dynamics, as they provide valuable insights into the distribution, recharge, and discharge of aquifers, as well as the interactions between groundwater and surface water systems. Seepage flow models are also vital in addressing issues related to water quality, soil moisture, and contaminant transport, which have significant implications for agriculture, water supply, and environmental protection.

The foundation of seepage flow models lies in Darcy's law, which relates the fluid velocity to the hydraulic conductivity and hydraulic gradient within the porous medium. Coupled with the principle of mass conservation, these models describe the flow of water through unsaturated and saturated zones, accounting for factors such as soil properties, heterogeneity, and anisotropy. Numerical methods, such as finite difference and finite element techniques, are often employed to solve the governing equations of seepage flow models, allowing for the accurate representation of complex subsurface conditions and boundary conditions.

12.1.1 Variational Formulation

The strong form of the seepage flow model needs to be solved,

$$\sigma + \nabla u = 0 \quad \text{in} \ \Omega \tag{12.1}$$

$$\nabla \cdot \sigma = f \quad \text{in} \ \Omega \tag{12.2}$$

subject to the following boundary condition:

© The Author(s), under exclusive license to Springer Nature Singapore Pte Ltd. 2024
A. Kumar and M. Saharia, *Python for Water and Environment*, Innovations in Sustainable
Technologies and Computing, https://doi.org/10.1007/978-981-99-9408-3_12

$$u = u_0 \quad \text{on } \partial D \tag{12.3}$$

$$-\sigma \cdot \hat{n} = g \quad \text{on } \partial N \tag{12.4}$$

where, the boundary, $\partial \Omega = \partial D \cup \partial N$. The corresponding weak form of Eqs. 12.1 and 12.2 can be given as

$$\langle \sigma, \tau \rangle + \langle \nabla u, \tau \rangle + \langle \sigma, \nabla v \rangle = -\langle \hat{n}, \sigma \rangle * v \tag{12.5}$$

In the above equations, σ represents the vector variable, and the velocity, and u represents the scalar variable, and pressure.

12.1.2 GMSH Code to Generate the Computational Domain

The given Gmsh code is used to describe a geometric 2D domain. Shown in Fig. 12.1 is a simple geometry consisting of points, lines, and surfaces.

First, it setups the geometry engine using the SetFactory ("OpenCASCADE") command. It then defines the point definitions—Points 1 and 2 represent the endpoints of a line segment, while points 3 to 9 define the vertices of a complex shape. Line definitions: Lines 1 to 10 connect the defined points, forming the boundaries of the geometry. They represent the edges of the shape.

Next, curve loops are defined—Curve Loop(1) defines a closed loop by specifying the lines that form the boundary. In this case, lines 1 to 10 are included. Plane Surface: Plane Surface(1) creates a 2D surface using Curve Loop(1). It represents the interior region enclosed by the loop.

Next, we define the physical curves and surfaces—Physical Curve and Physical Surface commands assign tags or labels to specific curves and surfaces. These tags can be used to define boundary conditions or apply material properties in subsequent simulations.

Therefore, this code defines a 2D geometry with specific points, lines, and surfaces. The defined surfaces and curves are given physical tags, which can be used to

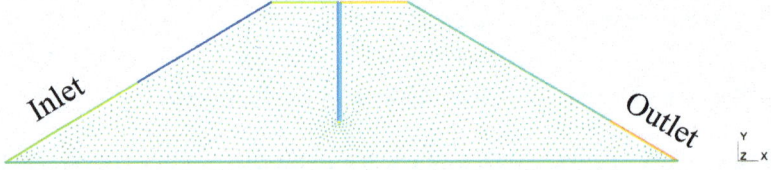

Fig. 12.1 Mesh as a result of the above given Gmsh code. The figure shows the inlet and the outlet. Pressure boundary conditions were applied to these boundaries. Rest all other boundaries were assigned a no-slip boundary condition. The generated mesh is listed in this Github link to Seepage Mesh

identify and assign specific properties or conditions to those regions during subsequent finite element simulations.

```
1   // Gmsh project created on Fri Jan  6 08:12:57 2023
2   SetFactory("OpenCASCADE");
3   //+
4   Point(1) = {0, 0, 0, 1.0};
5   //+
6   Point(2) = {50, 0, 0, 1.0};
7
8   Point(3) = {45, 3, 0, 1.0};
9   //+
10  Point(4) = {30, 12, 0, 1.0};
11  //+
12  Point(5) = {25.1, 12, 0, 1.0};
13  //+
14  Point(6) = {25.1, 3, 0, 1.0};
15  //+
16  Point(7) = {24.9, 3, 0, 1.0};
17  //+
18  Point(8) = {24.9, 12, 0, 1.0};
19  //+
20  Point(9) = {20, 12, 0, 1.0};
21
22  Point(10) = {10, 6, 0, 1.0};
23  //+
24  Point(11) = {0, 0, 0, 1.0};
25  //+
26  Line(1) = {1, 2};
27  //+
28  Line(2) = {2, 3};
29  //+
30  Line(3) = {3, 4};
31  //+
32  Line(4) = {4, 5};
33  //+
34  Line(5) = {5, 6};
35  //+
36  Line(6) = {6, 7};
37  //+
38  Line(7) = {7, 8};
39  //+
40  Line(8) = {8, 9};
41  //+
42  Line(9) = {9, 10};
43  //+
```

```
40   Line(10) = {10, 1};
41   //+
42   Curve Loop(1) = {9, 10, 1, 2, 3, 4, 5, 6, 7, 8};
43   //+
44   Plane Surface(1) = {1};
45   //+
46   Physical Curve("noslip", 11) = {1, 9, 8, 7, 5, 4, 3, 6};
47   //+
48   Physical Curve("inlet", 12) = {10};
49   //+
50   Physical Curve("outlet", 13) = {2};
51   //+
52   Physical Surface("dam", 14) = {1};
```

12.1.3 Python Code

The program simulates a steady seepage flow using the finite element method.

In lines 5–14, we import the necessary libraries, such as "numpy", "dolfinx", "pyvista", and "petsc4py", to facilitate the implementation of the finite element method and visualization.

In lines 17–22, the paths and names of the mesh file and the result file are specified to read the mesh and save the simulation results.

In lines 25–39, we define the mesh file, read using the "read_from_msh" function from "dolfinx". The domain, cell tags, and facet tags are obtained from the mesh file, which represents the computational domain and its geometric entities.

In lines 42–57, we define the finite element space and functions for the mixed element formulation. The "DRT" and "CG" elements are used to represent the velocity and pressure fields, respectively. The "W" function space is created to accommodate these mixed elements.

In lines 60–70, we define the trial and test functions that are derived from the mixed element space. The variational formulation of the seepage flow problem is defined using these trial and test functions.

In lines 73–105, we implement the boundary conditions for the Darcy flow simulation. Dirichlet boundary conditions are applied to the pressure variable at the inlet and outlet boundaries. The values of the pressure are assigned to the corresponding degrees of freedom.

In lines 108–129, we set up the solver where its parameters are configured using the PETSc options. The solver type is set to "preonly" with an LU preconditioner and the MUMPS solver for efficient solution of the linear system. The linear problem is solved using the specified solver configuration and boundary conditions. The solution is split into velocity and pressure components. The results are then visualized using the "PyVista" library. The pressure and velocity fields are projected onto suitable data structures, and the plots are displayed.

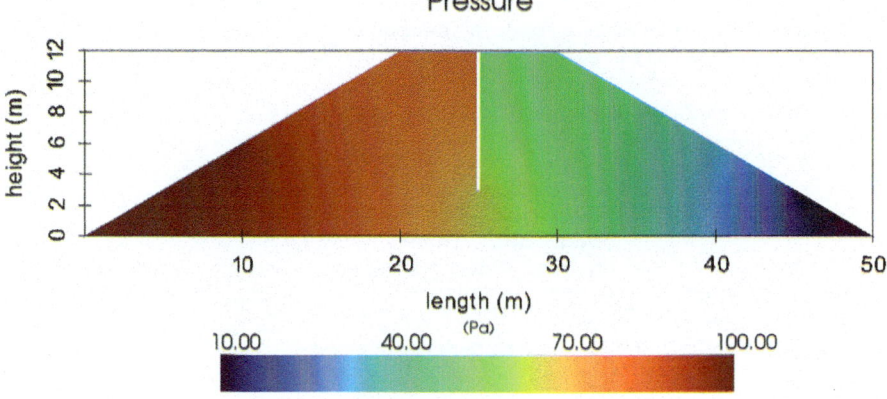

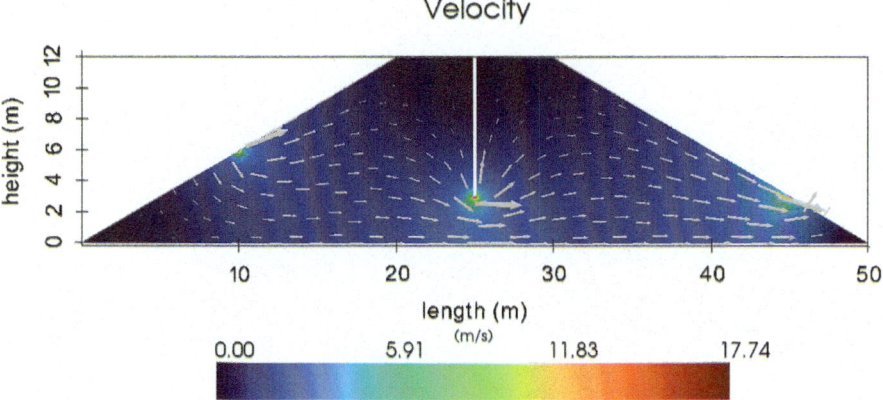

Fig. 12.2 Top: Pressure field; Bottom: Velocity field with arrows showing the flow direction. Worth noticing is that the model nicely captures the flow paths under the discontinuity

In lines 132–276, we set up the code where the simulation results for pressure and velocity are visualized using "PyVista". The plots are created using a "PyVista" plotter object, which allows for customization of the visualization parameters. The pressure and velocity fields are displayed in separate subplots, and appropriate color maps and scalar bars are added to enhance the visualization.

The resulting plots are saved to the specified result file path. Therefore, this code demonstrates the simulation of steady seepage flow using the finite element method.

```
1   """
2   Program to simulate a steady seepage
3   flow using finite element method.
4   """
5   import numpy as np
6   from dolfinx import fem, plot
7   from dolfinx.io.gmshio import read_from_msh
8   from ufl import (FiniteElement, MixedElement,
9                    TestFunctions, TrialFunctions,
10                   grad, dot, inner,
11                   ds, dx, FacetNormal)
12  from mpi4py import MPI
13  import pyvista as pv
14  from petsc4py.PETSc import ScalarType
15
16
17  """
18  File IO
19  """
20  simulationName = "Seepage_flow-2D"
21  meshName = "seepage_2D"
22  meshPath = "./meshes_gmsh/" + meshName + ".msh"
23
24
25  """
26  Computational domain - We load an external
27  mesh suitable for demonstrating seepage
28  flow
29  """
30  domain, cell_tags, facet_tags = read_from_msh(
31      filename=meshPath,
32      comm=MPI.COMM_WORLD,
33      rank=0,
34      gdim=2)
35
36  # Facet dimension where the boundary condition needs
37  # to be applied is 1 less than the domain's dimension
38  facetdim = domain.topology.dim - 1
39  n = FacetNormal(domain)
40
41
42  """
43  Finite element space and functions
44  """
45  # Define mixed elements for (velocity, pressure) pair
46  # that are stable (from literature)
```

```
38   DRT = FiniteElement(
39       family="DRT",
40       cell=domain.ufl_cell(),
41       degree=2)
42   CG = FiniteElement(
43       family="CG",
44       cell=domain.ufl_cell(),
45       degree=3)
46   W = fem.FunctionSpace(
47       mesh=domain,
48       element=MixedElement([DRT, CG]))
49
50
51   """
52   Trial, Test functions and
53   variational formulation
54   """
55   # Derive trial and test functions from the mixed element space
56   (sigma, u) = TrialFunctions(function_space=W)
57   (tau, v) = TestFunctions(function_space=W)
58   a = (dot(sigma, tau) +
59       dot(grad(u), tau) +
60       dot(sigma, grad(v))) * dx
61   L = -inner(n, sigma) * v * ds
62
63
64   """
65   Boundary conditions - this sets up the
66   necessary boundary conditions for the
67   Darcy flow simulation by assigning specific
68   pressure values at the inlet and outlet
69   boundaries, ensuring accurate modeling
70   of the flow behavior.
71   """
72   # Dirichlet boundary condition on pressure (u)
73   spc = 1   # signifies 2nd space where pressure variable dwells
74   Q, _ = W.sub(spc).collapse()
75   dof_inlet = fem.locate_dofs_topological(
76       V=(W.sub(spc), Q),
77       entity_dim=facetdim,
78       entities=facet_tags.find(12))
79   dof_outlet = fem.locate_dofs_topological(
80       V=(W.sub(spc), Q),
81       entity_dim=facetdim,
82       entities=facet_tags.find(13))
```

```python
70   p_inlet = fem.Function(Q)
71   p_inlet.interpolate(
72       lambda xd:
73       ScalarType(100.0) * np.ones((1, xd.shape[1])))
74   bc_pressure_inlet = fem.dirichletbc(
75       value=p_inlet, dofs=dof_inlet, V=W)
76
77   p_outlet = fem.Function(Q)
78   p_outlet.interpolate(
79       lambda xd:
80       ScalarType(10.0) * np.ones((1, xd.shape[1])))
81   bc_pressure_outlet = fem.dirichletbc(
82       value=p_outlet, dofs=dof_outlet, V=W)
83
84
85   """
86   Solver configuration - We create a linear
87   solver object for this model. The petsc
88   options configures the PETSc library to use
89   a LU-based preconditioner with MUMPS as the
90   underlying solver. These choices can significantly
91   impact the performance and accuracy of the numerical
92   computations performed using PETSc.
93   """
94   # Solve
95   problem = fem.petsc.LinearProblem(
96       a, L, bcs=[bc_pressure_inlet,
97                   bc_pressure_outlet],
98       petsc_options={
99           "ksp_type": "preonly",
100          "pc_type": "lu",
101          "pc_factor_mat_solver_type": "mumps"})
102
103  w_h = problem.solve()
104  velocity_h, pressure_h = w_h.split()
105
106  print("Solved! See the plots.")
107
108
109  """
110  Visualization - We use PyVista Plotter object
111  with a specific size and settings for line and
112  polygon smoothing. It then creates two subplots
113  within the plotter canvas to display the pressure
114  and velocity fields separately.
115  Then, for each subplot, the code projects the
116  simulation results onto a suitable PyVista data
117  structure, which consists of cells, cell types,
118  and coordinates. The projected solution is assigned
119  to the grid and visualized using different visualizations.
120  """
```

```
144  fontsize = 16
145  zoom = 1.6
146
147  # Setting pyvista plotter's backend
148  pv.set_plot_theme(pv.themes.DocumentTheme())
149  pv.set_jupyter_backend('None')
150
151  canvas_length, canvas_breadth = 1200, 1200
152
153  # Initialize a pyvista canvas with 1 row and 2 columns.
154  p = pv.Plotter(shape=(2, 1),
155                 window_size=(canvas_length, canvas_breadth),
156                 multi_samples=8,
157                 line_smoothing=True,
158                 polygon_smoothing=True, border=False)
159  # p.add_title("Darcy flow simulation")
160
161  # Project the solutions suitable with pyvista.
162  V0_h = fem.FunctionSpace(domain, ("CG", 2))
163  uh = fem.Function(V0_h, dtype=np.float64)
164  uh.interpolate(pressure_h)
165
166  # Get the cells, types and coordinates from the mesh.
167  cells, cell_types, x_ = plot.create_vtk_mesh(V0_h)
168
169  # Define an unstructured grid to store the data.
170  grid = pv.UnstructuredGrid(cells, cell_types, x_)
171
172  # Assign the projected solution onto the grid.
173  grid.point_data["u"] = uh.x.array.reshape(
174      x_.shape[0], V0_h.dofmap.index_map_bs)
175
176  # Select the 1st panel in the canvas.
177  p.subplot(0, 0)
178  p.add_text('Pressure',
179            position=(0.45, 0.75),
180            viewport=True,
181            shadow=True,
```

```python
182              font_size=fontsize)
183
184  # Set color map, zoom level, and other settings.
185  p.add_mesh(
186      mesh=grid.warp_by_scalar(
187          scalars="u", factor=0.0),
188      cmap="turbo",
189      scalar_bar_args={'title': "(Pa)",
190                       'label_font_size': fontsize + 8,
191                       'fmt': '%10.2f',
192                       'position_x': 0.25,
193                       'position_y': 0.01,
194                       'bold': False,
195                       'width': 0.5,
196                       'height': 0.2,
197                       'n_labels': 4}
198  )
199
200  # p.add_bounding_box()
201  p.show_bounds(ticks='both',
202                xlabel='length (m)',
203                ylabel='height (m)',
204                use_2d=True,
205                all_edges=True,
206                font_size=fontsize + 4)
207
208  # Display the XY plane
209  p.view_xy()
210
211  # Define the zoom level
212  p.camera.zoom(value=zoom)
213
214  V0_sigma = fem.VectorFunctionSpace(
215      domain, ("CG", 2))
216  usigma = fem.Function(
217      V0_sigma, dtype=np.float64)
218  usigma.interpolate(velocity_h)
219  cells, cell_types, x_ = plot.create_vtk_mesh(
220      V0_sigma)
221  grid = pv.UnstructuredGrid(cells, cell_types, x_)
```

```python
grid.point_data["u"] = uh.x.array.reshape(
    x_.shape[0], V0_h.dofmap.index_map_bs)

# Rearranging data for glyphs
points2d = usigma.x.array.reshape(
    x_.shape[0], V0_sigma.dofmap.index_map_bs)
points3d = np.hstack(
    [points2d, np.zeros([points2d.shape[0], 1])])
grid["vectors"] = points3d
grid.set_active_vectors("vectors")
glyphs = grid.glyph(
    scale=1, orient="vectors", tolerance=0.025,
    factor=0.5, geom=pv.Arrow().scale(
        xyz=(1, 1, 0.07), inplace=True))

p.subplot(1, 0)
p.add_mesh(mesh=glyphs,
           cmap="coolwarm", color="white",
           show_scalar_bar=False)

p.subplot(1, 0)
p.add_text('Velocity',
           position=(0.45, 0.75),
           viewport=True,
           shadow=True,
           font_size=fontsize)
p.add_mesh(
    mesh=grid.warp_by_scalar(
        scalars="u", factor=0.0),
    cmap="turbo",
    scalar_bar_args={'title': "(m/s)",
                     'label_font_size': fontsize + 8,
                     'fmt': '%10.2f',
                     'position_x': 0.25,
                     'position_y': 0.01,
                     'bold': False,
                     'width': 0.5,
                     'height': 0.2,
                     'n_labels': 4})
```

```
243   p.show_bounds(ticks='both',
244                   xlabel='length (m)',
245                   ylabel='height (m)',
246                   use_2d=True,
247                   all_edges=True,
248                   font_size=fontsize + 4,
249                   location='outer')
250
251   # Display the XY plane, set zoom level
252   # and show a title on the window
253   p.view_xy()
254   p.camera.zoom(value=zoom)
255   p.show(title="Seepage flow simulation")
256
257   print("Done!")
```

12.1.4   Discussion of the Results

Figure 12.2 shows the result of the seepage flow simulation. Due to the application of the pressure Boundary Conditions (BC) at the inlet and the outlet, the pressure decreases gradually from a high value to a low value. This BC is applied normally to the boundary at the two regions, viz. inlet and the outlet as seen in Fig. 12.2. The color gradation shows the variation. In the upper part, we see the pressure field. The high and low values are 100 Pa and 10 Pa, respectively. In the lower part of the figure, we see the velocity field. A quiver plot is overlaid on the velocity field that shows the flow of the fluid from a high-pressure to a low-pressure region. The discontinuity introduced at the center forces the fluid to have a higher velocity in the narrow region. The red color at the bottom edge gives an idea about the magnitude of the velocity. Additionally, we also see such properties at the top parts of the inlet and the outlet.

12.2   Groundwater Flow Model

Groundwater flow models are an essential component of subsurface hydrological investigations, designed to simulate and predict the movement and distribution of water within aquifers. These models play a crucial role in managing and protecting vital groundwater resources, which serve as a primary source of fresh water for millions of people worldwide. Groundwater flow models contribute to various applications, including water supply planning, contamination remediation, saltwater intrusion prevention, and the assessment of potential impacts from land-use changes, industrial activities, and climate change.

The fundamental principles of groundwater flow models are rooted in the mathematical representation of flow through porous media, such as DarcyÃ¢â‚¬â„¢s law, and the principle of mass conservation. The governing equations for groundwater flow are typically partial differential equations, accounting for parameters such as hydraulic conductivity, storage coefficients, and aquifer thickness. These equations are often solved using numerical methods, like finite difference or finite element techniques, to accurately represent complex aquifer systems and boundary conditions.

12.2.1 Variational Formulation

The strong form of the unsteady groundwater flow is given by the equation:

$$S\frac{\partial h}{\partial t} = K\nabla^2 h + f \quad \text{in } \Omega \tag{12.6}$$

subject to the boundary conditions:

$$u = u_D \quad \text{in } \partial D \times [0, T] \tag{12.7}$$

$$u = u_0 \quad \text{at } t = 0 \tag{12.8}$$

The corresponding weak form of Eq. 12.6 can be given as follows:

$$S\langle h - h_n, v\rangle - \Delta t K \langle \nabla h, \nabla v\rangle = \Delta t \langle f, v\rangle \tag{12.9}$$

Where, Ω is the computational domain, S is the storage, K is the permeability, h, h_n are the heights at the current time and previous time, respectively, and f is the source term. v is the test function. We solve for the unknown variable h as a function of time, t and space, x.

12.2.2 GMSH Code to Create the Computational Domain

This Gmsh code defines a 2D geometry using OpenCASCADE as the underlying geometry engine. It creates six points in 2D space, forming an L-shaped closed loop with lines connecting them. The points and lines are used to define a curve loop, representing the boundary of the geometry. The curve loop is then used to create a plane surface. The resulting surface is assigned a physical label "body" with tag 7, and the curves are assigned a physical label "8". This code sets up the geometry for further operations such as meshing or applying boundary conditions.

Fig. 12.3 Well simulation surface/plan view. Triangular mesh as a result of the above given Gmsh code. All the boundaries were assigned a no-slip boundary condition. Well locations (circles) are also shown on which the sink functions were applied. The generated mesh is listed in this Github link to L-shape Mesh

```
1   // Gmsh project created on Fri May 12 21:21:45 2023
2   SetFactory("OpenCASCADE");
3   //+
4   Point(1) = {0, 0, 0, 1.0};
5   //+
6   Point(2) = {10, 0, 0, 1.0};
7   //+
8   Point(3) = {10, 10, 0, 1.0};
9   //+
10  Point(4) = {7, 10, 0, 1.0};
11  //+
12  Point(5) = {7, 3, 0, 1.0};
13  //+
14  Point(6) = {0, 3, 0, 1.0};
15  //+
16  Line(1) = {1, 2};
17  //+
18  Line(2) = {2, 3};
19  //+
20  Line(3) = {3, 4};
21  //+
22  Line(4) = {4, 5};
23  //+
24  Line(5) = {5, 6};
25  //+
26  Line(6) = {6, 1};
27  //+
28  Curve Loop(1) = {5, 6, 1, 2, 3, 4};
29  //+
30  Plane Surface(1) = {1};
31  //+
32  Physical Surface("body", 7) = {1};
33  //+
34  Physical Curve(8) = {6, 5, 4, 3, 2, 1};
```

12.2.3 Python Code

This code solves the unsteady groundwater flow equation using the finite element method.

In lines 6–19, we import the necessary libraries, including "dolfinx", "pyvista", "numpy", and "petsc4py".

In lines 23–29, we define various parameters such as the file paths for mesh input/output, simulation time parameters, and computational domain details.

The program reads the mesh file using the "io.gmshio" module from "dolfinx" in 44 to 48.

The program creates the finite element space for the solution in lines 66–70. It also initializes the initial condition for the groundwater head using a user-defined function in lines 110–120.

In lines 123–132, we apply the boundary conditions which are specified using the "DirichletBC" function, which sets the groundwater head to zero on selected boundary facets.

In lines 142–148, the code also defines the forcing function for sink or "wells" in the domain.

In lines 154–165, the variational formulation of the groundwater flow equation is defined using trial and test functions, along with the sink term. The bilinear and linear forms are constructed using the "UFL" syntax.

Next, in lines 175–183, the solver configuration is set up, including the assembly of the matrix and creation of the linear solver using "PETSc". The solver type is set to LU factorization for efficient solving of the linear system.

In lines 76–107, the code sets up the plotting using "pyvista". It creates a grid based on the finite element space and defines a plotting function to visualize the groundwater head at different time steps.

We start the time-marching loop in line 190, where the program iterates over the specified number of time steps. It assembles the right-hand side vector, applies boundary conditions, and solves the linear system using the linear solver.

At every time step, the program calls the plotting function to visualize the groundwater head using "pyvista".

This code demonstrates the implementation of the finite element method for solving unsteady groundwater flow and provides a visualization of the evolving groundwater head over time.

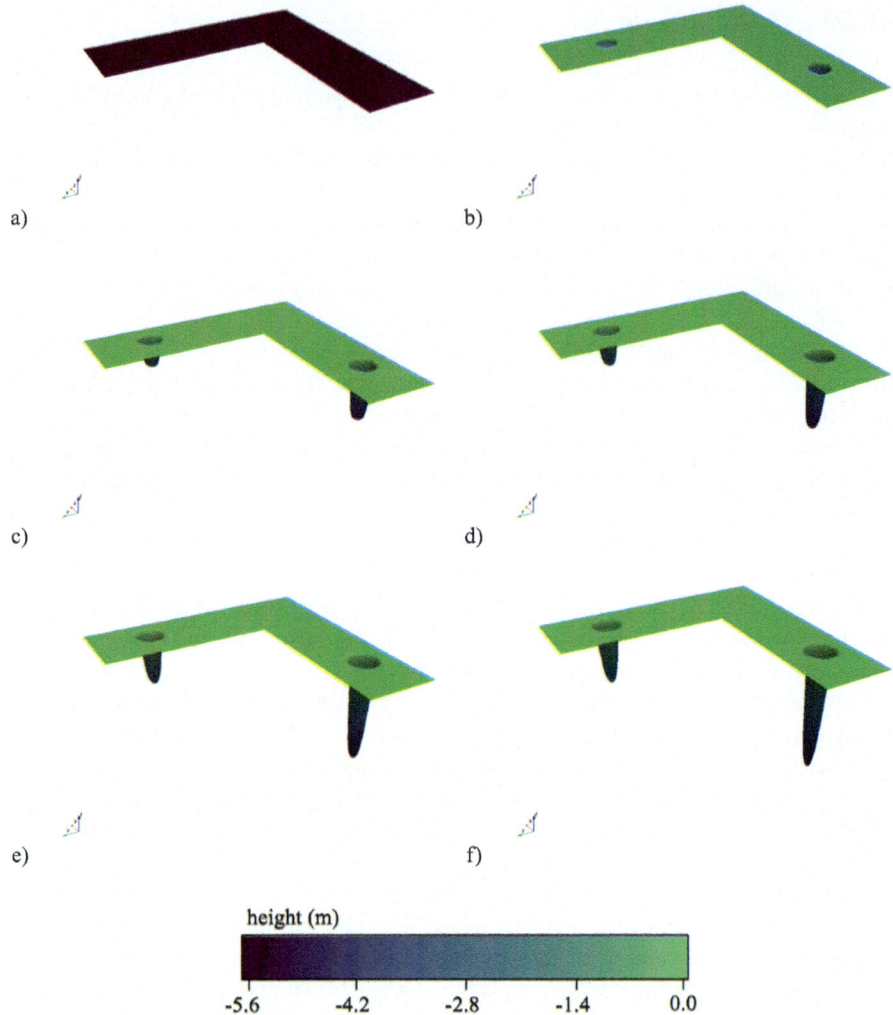

a)   b)

c)   d)

e)   f)

height (m)

-5.6   -4.2   -2.8   -1.4   0.0

Fig. 12.4 Groundwater flow simulation on an L-shaped region with two wells. From **a** to **f** Groundwater profile as a result of pumping at two wells with different depths. With the passage of time the water depth increases

```
1   """
2   Program to solve unsteady groundwater
3   flow equation using the finite-element
4   method.
5   """
6   import numpy as np
7   from dolfinx import plot
8   from dolfinx.fem import (FunctionSpace, Function,
9                            dirichletbc, locate_dofs_topological,
10                           form, petsc, Constant)
11  from dolfinx.io.gmshio import read_from_msh
12  from ufl import (FiniteElement, TestFunction, TrialFunction,
13                   grad, dot, dx, FacetNormal)
14  from mpi4py import MPI
15  import pyvista as pv
16  from petsc4py import PETSc
17  from petsc4py.PETSc import ScalarType
18  import matplotlib.pyplot as plt
19  from matplotlib.colors import LinearSegmentedColormap
20  pv.set_plot_theme(pv.themes.DocumentTheme())
21
22
23  """
24  File I/O
25  """
26  simulationName = "Groundwater_flow-2D"
27  meshName = "groundwater_2D"
28  meshPath = "./meshes_gmsh/" + meshName + ".msh"
29  resultPath = "./result/"
30
31
32  """
33  Running parameters
34  """
35  t_start = 0
36  t_stop = 1
37  Nt = 100
```

```
38  dt = (t_stop - t_start) / Nt
39
40
41  """
42  Computational domain
43  """
44  domain, cell_tags, facet_tags = read_from_msh(
45      filename=meshPath,
46      comm=MPI.COMM_WORLD,
47      rank=0,
48      gdim=2)
49
50  n = FacetNormal(domain)
51  surround = int(8)
52
53
54  """
55  Model parameters
56  """
57  S = Constant(
58      domain=domain, c=3.5)   # Storage
59  K = Constant(
60      domain=domain, c=0.05)   # Hydraulic conductivity
61
62
63  """
64  Finite element space
65  """
66  fe_elem = FiniteElement(family="CG",
67                          cell=domain.ufl_cell(),
68                          degree=2)
69  fe_space = FunctionSpace(mesh=domain,
70                           element=fe_elem)
71
72
73  """
74  Visualize
75  """
76  ocean = plt.cm.get_cmap("ocean")
77  ocean_modified = LinearSegmentedColormap.from_list(
78      "ocean_modified",
```

```
79      [ocean(i) for i in range(int(200), int(240))]])

81  def visualize_and_save(
82          h_, t, name_,
83          grid = pv.UnstructuredGrid(
84              *plot.create_vtk_mesh(fe_space))):

86      p = pv.Plotter(off_screen=True)

88      # Initialize grid values
89      grid.point_data[f"h({t})"] = h_.x.array.real

91      # Warp by values
92      warped = grid.warp_by_scalar(
93          scalars=f"h({t})", factor=1.0)

95      # Set angle and show
96      p.add_mesh(warped, cmap=ocean_modified,
97                  show_scalar_bar=False)
98      # p.show_grid()
99      p.show_axes()

101     # p.camera.view_angle = 15.0
102     p.camera.azimuth = 120
103     # p.camera.elevation = -15
104     p.camera.position = (-13.0, 15.0, 5.0)

106     p.show(title="Groundwater flow simulation",
107             screenshot=resultPath + name_)
108     p.close()

111 """
112 Initial condition
113 """
114 def initial_head(x):
115     return np.full(x.shape[1], 0.0)

118 h_n = Function(V=fe_space, name="h_initial")
119 h_n.interpolate(initial_head)
```

```
119  h_ = Function(V=fe_space, name="h_solution")
120  h_.interpolate(initial_head)
121
122
123  """
124  Boundary condition
125  """
126  fdim = domain.topology.dim - 1
127  bc = dirichletbc(value=ScalarType(0.0),
128                   dofs=locate_dofs_topological(
129                       V=fe_space,
130                       entity_dim=fdim,
131                       entities=facet_tags.find(surround)),
132                   V=fe_space)
133
134  # Store first field value
135  name_ = "h_{:.04f}_.png"
136  visualize_and_save(h_, t_start, name_.format(t_start))
137
138
139  """
140  Forcing function
141  """
142  def sink(x):
143      vals = np.full(x.shape[1], 0.0)
144      well_1 = (x[0] - 2.5) ** 2 + (x[1] - 1.25) ** 2 < 0.15
145      well_2 = (x[0] - 8.75) ** 2 + (x[1] - 7.5) ** 2 < 0.15
146      vals[well_1] = -20.0   # deeper water level
147      vals[well_2] = -10.0   # shallower water level
148      return vals
149
150
151  """
152  Variational formulation
153  """
154  # Trial and Test functions
155  h = TrialFunction(function_space=fe_space)
156  v = TestFunction(function_space=fe_space)
157
158  # Source function
159  depth_fun = Function(V=fe_space)
```

```
160    depth_fun.interpolate(lambda x: sink(x))
161
162    # Variational form
163    a = S*h * v * dx
164    a += dt*K * dot(grad(h), grad(v)) * dx
165    L = (S*h_n + dt*depth_fun) * v * dx
166
167    # Bilinear and linear forms
168    bil_form, lin_form = form(a), form(L)
169
170
171    """
172    Solver config
173    """
174    # assemble
175    A = petsc.assemble_matrix(bil_form, bcs=[bc])
176    b = petsc.create_vector(lin_form)
177    A.assemble()
178
179    # solver
180    linear_solver = PETSc.KSP().create(domain.comm)
181    linear_solver.setOperators(A)
182
183    linear_solver.setType(PETSc.KSP.Type.PREONLY)
184    linear_solver.getPC().setType(PETSc.PC.Type.LU)
185
186
187    """
188    Time-marching through solution
189    """
190    t = t_start
191    for i in range(Nt):
192
193        # Update time
194        t = t + dt
195
196        # Set all the entries of the local vector b to zero
197        # and then assemble the global vector linear_form
198        # using the values from the local vector.
199        with b.localForm() as loc_b:
200            loc_b.set(0)
```

```
200      petsc.assemble_vector(b, lin_form)
201
202      # Apply boundary conditions to the right-hand side
203      # vector 'b' of the linear system by modifying its
204      # values according to the specified boundary conditions.
205      # The 'ghostUpdate' operation ensures that the values of
206      # 'b' are correctly communicated and updated across
207      # parallel processes.
208      petsc.apply_lifting(b, [bil_form], [[bc]])
209      b.ghostUpdate(
210          addv=PETSc.InsertMode.ADD_VALUES,
211          mode=PETSc.ScatterMode.REVERSE)
212      petsc.set_bc(b, [bc])
213
214      # Solves a linear system of equations represented by the
215      # matrix equation Ax = b. Solves the system
216      # and updates the solution vector h_.vector. The
217      # h_.x.scatter_forward() operation distributes the
218      # updated values of h_ across the distributed mesh for
219      # further processing or visualization.
220      linear_solver.solve(b, h_.vector)
221      h_.x.scatter_forward()
222
223      # Assign previous to the current
224      h_n.x.array[:] = h_.x.array
225
226      # Store subsequent field values
227      name_ = "h_{:.04f}_.png"
228      if i % 1 == 0:
229          visualize_and_save(h_, t, name_.format(t))
230
231      print("Done step {}".format(t))
232
233  print("Done!")
```

12.2.4 Discussion of the Results

Figure 12.4 shows the results of the unsteady groundwater simulation on a computational domain as shown in Fig. 12.3. All the boundaries were applied with a Dirichlet boundary condition on the height variable (the unknown). The two circular regions depict the wells where the source (sink) functions were applied. The circular wells were realized by applying the equation of the circle. The magnitudes of the sinks were different at the two wells where the top well (in Fig. 12.3) had a lower sink magnitude and the bottom well (in Fig. 12.3) had a higher sink magnitude. The unsteady groundwater flow equation was solved in 100 steps. With each step, we can see that the water level profile gradually differs for the two wells. This shows that wells with unequal bottoms may induce different sink strengths resulting in a variation in the water profile.

Part IV
Environmental Applications

Chapter 13
Transport Phenomena

Water contamination is a major concern in modern urban setups. It is a result of the mixing of a toxic substance, liquid or solid, with water that renders the water unusable for most purposes. In many scenarios, contamination happens accidentally, although intentional cases cannot be denied. Once released into the environment the contaminated water finds its way to the water table via different modes of transport. The contaminant may undergo a number of chemical changes on the way through the unsaturated zones before reaching the aquifer. The aquifer then helps transmit the liquid contaminants toward its outlets, and in the process, its concentration reduces gradually due to various physical, biological, and chemical reactions. This water that is found in springs, lakes, and rivers needs prior treatment for any use. Therefore, it is very important that the quality and quantity of these water resources be managed appropriately. Perhaps, technology is the most obvious means for the prevention and management of water contamination. Not to mention, legislation and education also play a significant role. However, it is the management of contamination that has to play a bigger role because pollution continues to happen.

Consider an example of leaching where the contaminants, after getting released from a point source, or a storage tank, percolate through the unsaturated zones within the subsurface until they reach the water table. The groundwater flow then further carries the polluted water to other outlets. This chapter is dedicated to laying out the mathematical background of contaminant transport in the subsurface. These contaminants are present in the dissolved state and their concentration determines the water quality. The change in concentration is a direct manifestation of chemical reaction, precipitation, and dissolution of solid contaminant. Such changes are required to be put in the governing partial differential equation where they often form a part of the forcing function/term. It is to be noted that we work with macroscopic models that can predict the flow/transport of the chemical species (often reactive) when the carrier, here water, flows through the subsurface. The contaminant transport models differ from the single/multiphase flow models in terms of the dependent variable being modeled. While it is the mass density for the fluid flow models, it is the concentration of the chemical species for the contaminant transport models.

We begin by first describing a few important terminologies relevant to contaminant transport models. There exist many measures of the quantity of the participating chemical species. Perhaps, the most popular among them is "concentration". Others are "mass concentration", "molar concentration", "molar fraction", and the "equivalent concentration". The concentration of a chemical species can be understood as the amount of the species, often measured in *gram* or the *mole* per unit volume of the intensive phase, the water, in our case.

Mass concentration is often expressed as *grams per unit liter* or *g/L*. A more popular unit to specify the quality of water is the *mg/L*. The molar concentration is expressed as the number of *moles per unit liter*, often written as *mol/L*. The molar fraction is defined as the ratio of the number of moles of the component to the total number of moles of the intensive phase. The measure is relevant for models where there is an interaction among the participating species, i.e., the components that undergo chemical reactions. The equivalent fraction is defined as the amount of the participating species that can react with the specified amount of another species/substance in a specific reaction.

13.1 Contaminant Transport Processes

Contaminant transport in the environment is a continuous phenomenon that refers to the movement of contaminants and pollutants or any harmful substances via the water network in a given area. One of the most important issues that we face today is groundwater contamination. It is not uncommon when the water in wells, rivers, and streams is affected by the mixing of contaminants that may gradually escape into the environment after a prolonged application of pesticides, industrial wastes/chemicals, and fertilizers present on the ground surface, land, or seep through the soil from underground storage tanks. We often find reports of water systems being polluted by agricultural by-products, sewage, and chemicals. Most of the time, erosion and runoff are to be blamed for such contamination. Yet another example is leachate transport where water percolates through waste disposal sites. Often, there is a high amount of heavy metals, pathogens, and other organic compounds that are carried with the water, which when traveling to the groundwater system can affect the resources dependent on it. Soil erosion is also a major source of contaminant transport. Agricultural lands and construction sites contribute to erosion which consequently increases sedimentation in the close by water systems. Another example is the aquatic ecosystems where harmful elements such as mercury and toxic polychlorinated biphenyls or PCBs accumulate in aquatic animals and fishes, consequently posing a health threat to humans.

Contaminant transport is a manifestation of three fundamental processes—diffusion, advection, and reaction. The contaminants are usually reactive in nature which brings about physical and chemical changes as the contaminant is transported from one place to another. Transportation via diffusion occurs when there is a difference between the concentration of the contaminants. Diffusion is a gradual process due to the random motion of the fluid particles. Advection is the process by which the contaminants are carried with the flow of the water through which it flows. Compared to diffusion and reaction, transport via advection is relatively fast.

The overall behavior of contaminant transport is a manifestation of these three processes and consequently is modeled using differential equations that consider these mechanisms and mutual interactions. In the upcoming sections, we describe different models that can be used with contaminant transport.

13.2 1D Diffusion Equation

In hydrology, the diffusion equation is used to describe the spread of contaminants or pollutants in groundwater. The equation models the migration of the contaminant over time in the subsurface due to molecular diffusion. The diffusion coefficient determines the rate at which the substance spreads, while the Laplacian operator characterizes the spatial gradient of the concentration. The 1D diffusion equation can be expressed as

$$\frac{\partial [u]}{\partial t} = D \frac{\partial^2 [u]}{\partial x^2} \tag{13.1}$$

where, $[u]$ is the concentration of the substance, t is time, D is the diffusion coefficient, and $\frac{\partial^2}{\partial x^2}$ is the Laplacian Δ operator. The equation describes the rate of change of the contaminant's concentration with respect to the product of the diffusion coefficient D and double spatial derivative of the concentration. Physically, D states the ease with which the pollutant can move through a medium. Determination of this quantity is often a complex task but due to its importance, researchers and scientists continue to devise new methods for estimating its value. Some of the reasons that make it difficult to estimate the coefficient's value are:

- The occurrence of more than one contaminant, each having a different value of coefficient, can complicate the determination of a single value of the diffusion coefficient.
- Generally, the environment where contaminant travels is mostly heterogeneous due to varying porosity, permeability, and tortuosity. These non-constant factors can still complicate the determination.
- The values of the factors can change over time making determining the diffusion coefficient difficult.

By solving the diffusion equation, hydrologists are able to forecast the movement and distribution of contaminants in groundwater over time, providing valuable information for assessing risks to water resources and designing remediation plans. The equation is also used in other hydrologic contexts, such as modeling the transport of pollutants in surface water or the transfer of heat in soils.

13.3   1D Diffusion-Reaction Equation

Reaction in the context of contaminant transport means any change in the concentration of the contaminant over time. When combined with the diffusion equation 13.1 the system reads

$$\frac{\partial[u]}{\partial t} = D\frac{\partial^2[u]}{\partial x^2} + R[u] \tag{13.2}$$

where R is called the rate of reaction with the rest having the same meanings as the diffusion equation. The reactions occur between the contaminant and the environment where they travel and they can be classified as follows:

- In groundwater, oxidation/reduction reaction is an important process that decides the fate of the contaminant where there is a deficiency of dissolved oxygen.
- Degradation of the organic compounds by micro-organisms such as bacteria and algae is also a common subsurface phenomenon. These organisms help break down more complex contaminants into simple compounds.
- Sorption is another process that affects the movement of the contaminant into the environment. It is a process by which the contaminants are adsorbed onto the surface of the medium.
- It is also common that one contaminant can get converted to another via some biological and chemical process. This generally results in the formation of new compounds with significantly new properties.

Understanding the properties of these reactions is essential as it can help conduct risk assessment and prepare remediation techniques, once the contaminant is dispersed into the environment. An accurate modeling of the reaction process can also help develop management strategies.

13.4   1D Advection-Diffusion Equation

In hydrology, advection-dominated flow refers to a way that water and other substances move in a groundwater or surface water system. Instead of spreading out and mixing with their surroundings, these substances are carried along in a cohesive unit as the water moves. This type of flow is typically found in places like aquifers, rivers, and streams, and can greatly impact the availability and quality of water. A one-dimensional advection-diffusion model can be stated as

$$\frac{\partial[u]}{\partial t} + w\frac{\partial[u]}{\partial x} = D\frac{\partial^2[u]}{\partial x^2} \tag{13.3}$$

where w is the advection coefficient having the unit of velocity with others having the same meaning as the diffusion case.

Advection dominated flow is important for flood control because it determines the speed and direction of the movement of water. If this type of flow is strong, water can quickly flow downstream, which can help to prevent or reduce the damage caused by floods. The reason is that the water does not stay in one place for very long and cannot cause as much damage. Understanding advection-dominated flow can help make better decisions for flood control, such as building better structures or predicting how rivers will behave during floods. In short, considering advection-dominated flow can help protect communities and infrastructure from flood damage.

13.5 A 2D Simulation Using Navier-Stokes Equations

Understanding of the fluid flow phenomenon is essential as it can affect many aspects of human life, such as flood control and water resources. Computational fluid dynamics, or CFD, is a technological advancement that helps us define the governing equations of fluid flow. The equations are widely known as the Navier-Stokes equations, which represent the mathematical models describing fluid flow behavior and its interactions in riverbeds, rocks, and debris.

The Navier-Stokes equations are specified using a set of partial differential equations that are solved for velocity and pressure. These equations are essential for understanding and predicting fluid flow behavior in rivers and other water bodies. They provide a general system of PDEs that, when combined with the physical properties of the fluid, say water, can be used to model its movement in rivers and to predict pressure, velocity, and other derived quantities such as drag and lift. However, the latter are more relevant in aerodynamics and atmospheric flows.

Another critical equation in the analysis of flow in rivers is the continuity equation. It states that the amount of fluid entering a system must equal the amount of fluid exiting it. It facilitates the balance between the volume of water entering the river system and the volume leaving it, thereby helping predict water flow over time. These equations play a significant role in understanding the behavior of a river system in events such as flood and change in the physical properties of the constituent fluids. Therefore, a numerical simulator that solves the Navier-Stokes equations qualifies to become an essential tool for hydrologists and water resource managers as it helps them make decisions about flood control and mitigation.

The Navier-Stokes equations can be understood as Newton's second law of motion for fluids. The equations consist of two parts. Equation 10.1 is restated here as Eq. 13.4 with the addition of the continuity part. In this section, we describe a numerical model to solve the incompressible Navier-Stokes equations for single-phase flow. The word "incompressible" means that the density of the fluid does not change with the application of external pressure. This also means that the volume of the water (fluid) remains constant as it flows along. Hence, in the second part of Eq. 13.4 we drop the $\frac{\partial \rho}{\partial t}$ term from the LHS. The inertial forces are represented by the terms o the LHS; then the first term in the RHS consists of the pressure and viscous forces as given by Eq. 10.9's third part and Eq. 10.12. The term **f** represents all external forces.

$$\rho \left(\frac{\partial \mathbf{v}}{\partial t} + (\mathbf{v} \cdot \nabla \mathbf{v}) \right) = \nabla \cdot \sigma + \mathbf{f}$$

$$\nabla \cdot \mathbf{v} = 0$$

(13.4)

The governing equations define the motion of fluids when subjected to different boundary conditions. In these equations ρ stands for the density of the fluid, μ is the dynamic viscosity, $\mathbf{v}$, the velocity vector, σ, the Cauchy stress tensor, $\mathbf{f}$ stands for the external force at the source or sink, and p denotes the pressure. There exists a limited number of analytical solutions for the equations. To date, the equations are solved using numerical schemes, and there is a vast literature on the challenges and their possible workarounds.

Some of the reasons for their non-triviality are: the equations involve many variables like pressure, velocity, and viscosity among other factors, which make them difficult to solve. The interplay between these variables makes the system complex. Secondly, the equations are highly nonlinear. The conventional matrix solvers require quite an amount of tuning for the given system to solve. Thirdly, the Navier-Stokes equations are plagued with non-uniqueness, i.e. for a given set of boundary conditions there may exist multiple solutions that would fit the observed data. In addition to all of these, the equations generally require large processing power as the computational requirements are high, even for a simple 3D problem.

In fluid flow modeling using Navier-Stokes equations, high Reynolds numbers are indicative of turbulent flows. This is true for river systems where small-scale motions occur which results in chaotic behavior. Regarding the stability of the solutions, there exist multiple schemes as cited by the literature. Here, we give a very brief explanation of this scheme and adapt it to the Fenics program illustrated next.

The Crank-Nicolson method is a commonly used numerical technique for solving partial differential equations (PDEs) in both temporal and spatial domains. It is known for its balance of stability and accuracy, which results in second-order precision. The method is frequently employed in problems concerning heat transfer and fluid dynamics, among others. As an implicit method, the solution at each time step depends on the solution at the subsequent time step, thus requiring the solution of a linear system of equations.

The method consists of three main steps:

1. The first step is called the predictor step. It consists of solving the momentum equation, resulting in the system being progressed to a mid-time-step position.
2. Following the predictor step, an initial projection may be carried out to ensure that the velocity field at the mid-time-step is divergence-free.
3. The algorithm then performs a corrector step, which uses the time-centered estimates of velocity to attain the final state at the current time step.

13.5.1 *Construction of the Computational Domain*

Any finite element program relies on a quality mesh for an analysis. The requirements of the mesh are quite strict when compared to other numerical methods. In order to generate a good quality mesh, we use the GMSH open-source software.

Figure 13.1 is representative of an aquifer with a surface area of ≈1750 m ^{2}. We introduce the inlet and the outlet as shown in Fig. 13.1. The inlet has a width of 5 m and the outlet has a 10 m wide opening. Two obstacles are introduced which represent islands amidst the aquifer.

Once the geometry of the domain is created, we specify markers to specific boundaries in the domain. The different markers are shown in different colors in Fig. 13.2. The inlet and outlet are in navy blue and purple colors. The obstacles/islands and the rest of the boundaries are shown in green and pink colors respectively. After the specification of the markers, the remaining task is meshing. The software provides suitable options to export the "demarcated" mesh in FEniCS readable formats. The native mesh format has a ".msh" extension—this is readily parsed by the FEniCS's "read_from_mesh" function.

In order to create the computational domain we use the GMSH open-source software. Referring to Fig. 13.1, the GMSH code required to generate the domain is given in Sect. 13.5.2. The first line in the code "SetFactory("OpenCASCADE")" is specified to set up the geometry kernel. Line 3 sets the value of the parameter "h". The value of 12.0 was chosen after some trials where the criteria were to coarsen

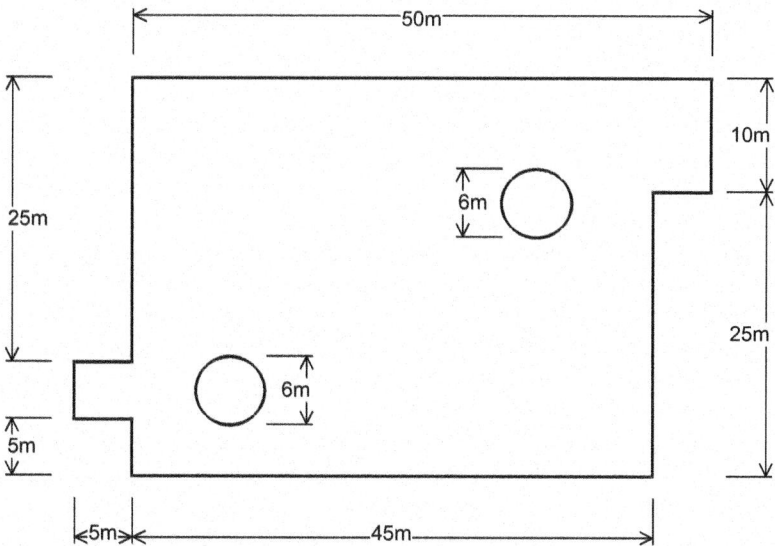

Fig. 13.1 Computational domain generated using GMSH. The code to generate the geometry is listed in this Github link to Aquifer Geometry

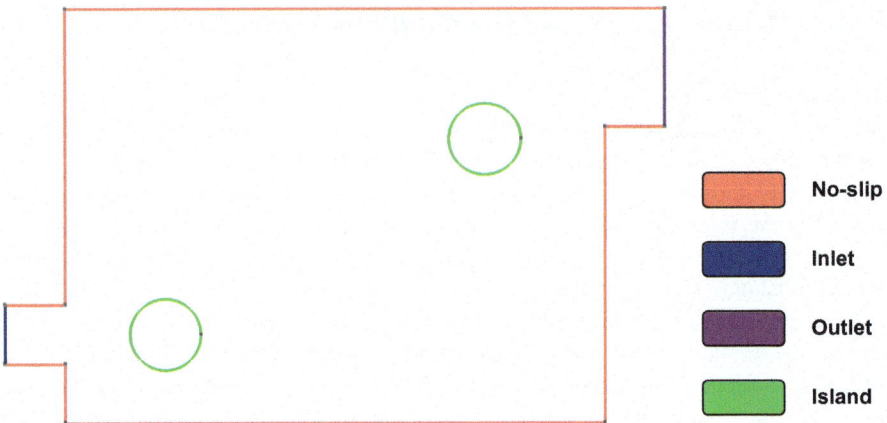

Fig. 13.2 Computational domain with designated markers. The legend shows the boundary condition to be associated with each marker. We intend to simulate a fluid flow from the inlet to the outlet. The circular obstacles are called islands and they cause obstruction to the flow, making the visualizations more appealing

the elements in the inlet and the outlet. Figure 13.3 is the result of meshing with this value. A value less than 12.0 resulted in the elements being too fine.

The keywords "Point" specify the coordinates of the point in the 2D plane where the plane is realized by setting the third entry as zero. The last entry is a term containing the parameter "h". We note that points 1, 2 and 11, 12 have this entry three times larger than others. These points correspond to the inlet and outlet and hence a larger parameter value ensures large element sizes. But still looking at Fig. 13.3 one can notice fine elements being present around the inlet, outlet, and the obstacles. The presence of fine elements helps achieve better accuracy in these regions. We intentionally settle for this mesh so as to capture the phenomena such as vortex street due to the no-slip boundary condition. The regions with larger elements are expected to exhibit homogeneous patterns.

The keyword "Line" is used to specify the line segment defined using a pair of points. There are 10 such line segments that form a closed boundary of the domain. The obstacles are realized using "Circle" keywords where the first three entries specify a point in the circumference, followed by an entry of the radius. The last two entries help specify the completeness of the arc; a value of "2π" denotes a complete closed circle.

The "Curve Loop" keyword works with closed boundaries only and has been used to identify the three closed loops—the outer boundary, and the inner two obstacles. The "Plane Surface" keyword is used to specify the surface excluding the obstacles. The physical groups within GMSH help to specify the boundary markers. As can be seen in lines 34–38, the different boundaries have been given suitable names along with an integer value. Although, either of these can be used with boundary conditions, FEniCS/DOLFINx mandates using the integer values for the specification. The sets

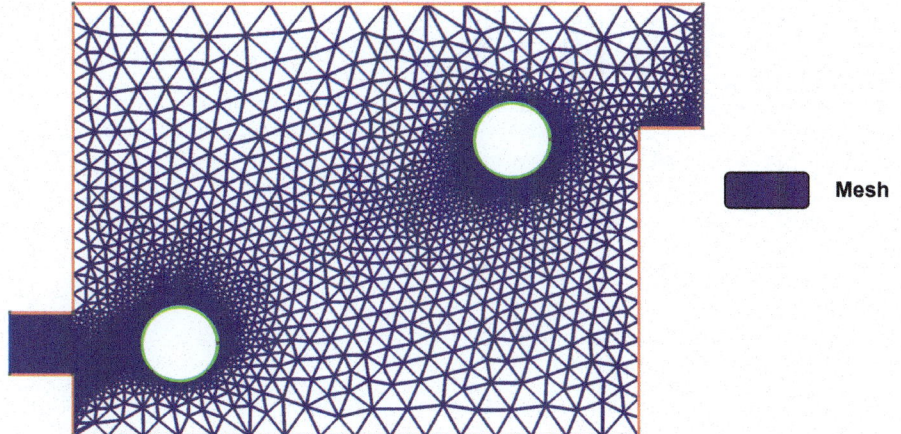

Fig. 13.3 Triangular mesh elements were generated for the domain. Mesh resolution has been increased near the islands (circles), the inlet, and the outlet. The generated mesh is listed in this Github link to Aquifer Mesh

in the RHS denote the line segments that a particular physical curve must refer to. In the last line, we also define a physical surface called "aquifer" for the sake of completeness.

Once the markers are set we mesh the domain with triangular mesh elements as shown in Fig. 13.3. The "*h*" parameter controls the density of the elements near the points. Additionally, three more iterations comprising of "Refine by splitting" and "Recombine 2D" were also carried out; the shown mesh is a result of all those operations. These options are only available in the graphical user interface of the GMSH software.

13.5.2   GMSH Code to Generate the Computational Domain

```
1   SetFactory("OpenCASCADE");
2   //+
3   h=12.0;
4   Point(1) = {0, 0.50, 0, h*3};
5   Point(2) = {0.50, 0.50, 0, h*3};
6   Point(3) = {0.50, 0, 0, h};
7   Point(4) = {5.00, 0, 0, h};
8   Point(5) = {5.00, 2.50, 0, h};
9   Point(6) = {5.50, 2.50, 0, h};
```

```
10   Point(7) = {5.50, 3.50, 0, h};
11   Point(10) = {0.50, 3.50, 0, h};
12   Point(11) = {0.50, 1.00, 0, h*3};
13   Point(12) = {0, 1.00, 0, h*3};
14   //+
15   Line(1) = {1, 2};
16   Line(2) = {2, 3};
17   Line(3) = {3, 4};
18   Line(4) = {4, 5};
19   Line(5) = {5, 6};
20   Line(6) = {6, 7};
21   Line(7) = {7, 10};
22   Line(8) = {10, 11};
23   Line(9) = {11, 12};
24   Line(10) = {12, 1};
25   //+
26   Circle(11) = {1.35, 0.75, 0.0, 0.3, 0, 2*Pi};
27   Circle(12) = {4.0, 2.4, 0.0, 0.3, 0, 2*Pi};
28   //+
29   Curve Loop(1) = {3, 4, 5, 6, 7, 8, 9, 10, 1, 2};
30   Curve Loop(2) = {11};
31   Curve Loop(3) = {12};
32   Plane Surface(1) = {1, 2, 3};
33   //+
34   Physical Curve("inlet", 13) = {10};
35   Physical Curve("outlet", 14) = {6};
36   Physical Curve("noflow", 15) = {9, 8, 7, 5, 4, 3, 2, 1};
37   Physical Curve("island", 16) = {11, 12};
38   Physical Surface("aquifer", 17) = {1};
```

13.5.3 Variational Formulation

The variational formulation of the Navier-Stokes equation consists of three parts. The first part can be given as

$$
\begin{aligned}
&\int_\Omega \frac{\rho_w}{dt} \langle u - u_n, v \rangle dx \\
&+ \int_\Omega \langle \langle \tfrac{3}{2} u_n - \tfrac{1}{2} u_{n1}, \tfrac{1}{2} \nabla(u + u_n) \rangle, v \rangle dx \\
&+ \int_\Omega \tfrac{1}{2} \mu \langle \nabla(u + u_n), \nabla v \rangle dx \\
&- \int_\Omega \langle p, \nabla \cdot v \rangle dx \\
&+ \int_\Omega \langle f, v \rangle dx = 0
\end{aligned}
\tag{13.5}
$$

where, ρ_w is the density of water, u denotes the velocity at the current step. u_n, u_{n1} denote the first and second steps from the current step in the past within a time loop. v denotes the test function, p is the pressure and f is the source function.

This formulation uses an Adams-Bashforth of degree 2 based time discretization technique.

The second part of the variational form can be given as

$$\int_{\Omega} \langle \nabla p_-, \nabla q \rangle \, dx - \int_{\Omega} \frac{\rho_w}{dt} \langle \nabla \cdot u_s, q \rangle \, dx = 0 \tag{13.6}$$

where, p_- is a function corresponding to intermediate pressure; q is the test function from the pressure's function space.

The third part of the variational form can be given as

$$\int_{\Omega} \rho_w (\langle u_s, v \rangle - \langle u, v \rangle) dx = \int_{\Omega} \frac{1}{dt} \langle \nabla \phi, v \rangle \, dx \tag{13.7}$$

where, ϕ is a function defined on the pressure's function space.

13.5.4   Python Code

In this section, we describe the code used to do the Navier-Stokes simulation for incompressible fluid flow. From lines 1–18, we first import all the necessary libraries.

In lines 20–29, we specify the name of the simulation, the location of the externally generated mesh file, and the location where the results would be stored.

In line 31, we check the presence of the mesh file with an "if" conditional statement. Once the mesh is found at the designated location we load the mesh file along with the cell tags and facet tags. This is done in lines 39–43.

In lines 45–49, we assign appropriate integers to the different boundary markers. These values were noted while creating the mesh in the GMSH mesh generation software. In line 52, we specify the dimension of the facet which is one less than the dimension of the domain.

From lines 55–65, we specify the properties of the fluid like the kinematic viscosity and fluid density. In lines 66–71, the time loop parameters like the start time, the end time, Δt, number of steps between the start and end time are specified. We also convert the Δt into a "PETSc" data type to enable changes when required.

In lines 73–86, we specify the finite element types and the function spaces. "Vspace" is a vector space for the velocity variable; "Qspace" is a scalar space for the pressure variable.

From lines 89–107, we write a class definition to realize a time-dependent boundary condition; the "__init__" method is used to initialize the value of time, t; the "__call__" method is used to compute a time-dependent velocity. Because our inlet is parallel to the X-axis, we only modify the values of the first row in lines 104–106. The profile can be seen to vary with the Y-coordinate as well.

Next, we compute the boundary conditions (BCs) from lines 110–157. There are 4 blocks of code for the BCs at the inlet, the walls, the islands, and the outlet. We

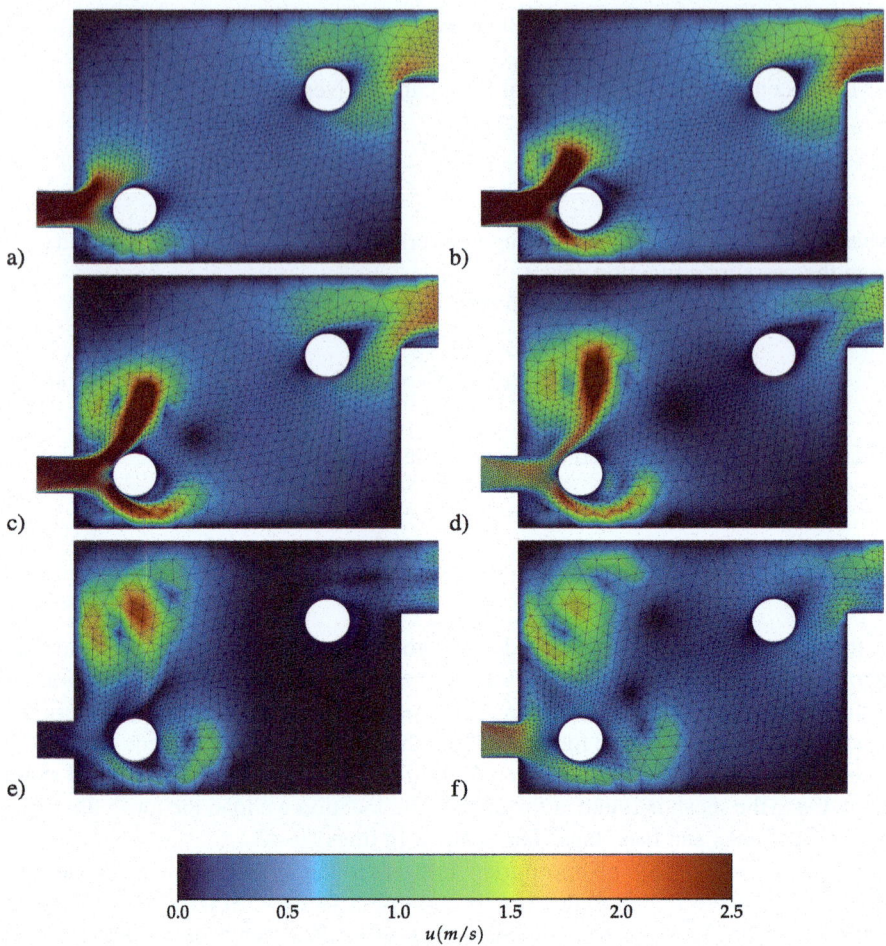

Fig. 13.4 Velocity field as computed by the above program Sect. 13.5.4 at 6 different time, $t = 0.03$ s, 0.33 s, 1.0 s, 1.66 s, 2.33 s, 3.00 s. One can notice specific patterns that emerge due to the presence of the two circular-shaped islands imposing the no-slip boundary condition along with the surrounding boundaries

apply a Dirichlet BC for the velocity variable at the inlet, outlet, and the islands. We specify a Dirichlet condition on the pressure variable at the outlet.

In lines 159–180, we specify the trial and test functions, the functions required for intermediate steps of the procedure. "u_, p_" stand for the functions defined on the "Vspace" and "Qspace" for velocity and pressure variables, respectively. "u, v" are trial and test functions defined on the "Vspace". Similarly, "p, q" are trial and test functions defined on the "Qspace". "ϕ" is another function defined on the "Qspace". The reader is referred to the variational formulation for a detailed description. In lines 182–185, a vector source function is defined.

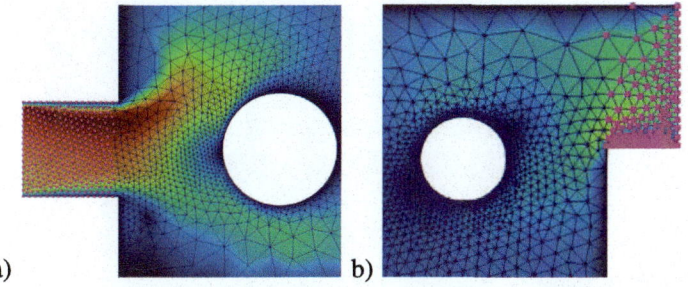

Fig. 13.5 Inlet and outlet velocity probes: **a** Pink dots show the nodes that were used to record the velocity data at the inlet; **b** Pink dots show the nodes that were used to record the velocity data at the outlet

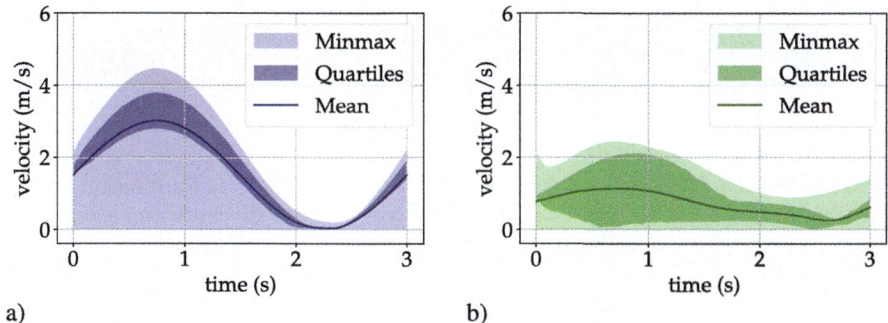

Fig. 13.6 Inlet and outlet velocity data statistics. The inside color bands indicate the 1st and 3rd quartiles. The outer color bands indicate the minimum and maximum of the velocity recorded at the nodes. The selected nodes are shown in Fig. 13.5

In lines 187–203, the weak form of the first step is defined. In lines 205–208, we use the "lhs" and "rhs" methods of the "dolfinx" library to extract the bilinear and linear forms from the weak form expression above. These methods come in handy when the weak forms are lengthy.

In lines 210–213, we specify the weak form of the second step. In lines 215–218, we specify the weak form of the third step.

In lines 220–234, we assemble the matrix and the vector that are to be solved in the "PETSc" solver. In lines 236–253, we set up separate solvers for the three steps of the Navier-Stokes simulation.

In lines 255–272, the file IO operation objects are specified; we write two solution files for the velocity and pressure fields. The (file_mode= "w") specifies a write operation; the encoding is specified as hierarchical data format or HDF5. For both the solution files we first write the meshes and the initial time.

In line 274, a variable with the shape as that of the "u_" is initialized.

In lines 276–372, we implement the three-step procedure for the simulation. We begin by initializing a "with" context manager writing the velocity data file. A dataset

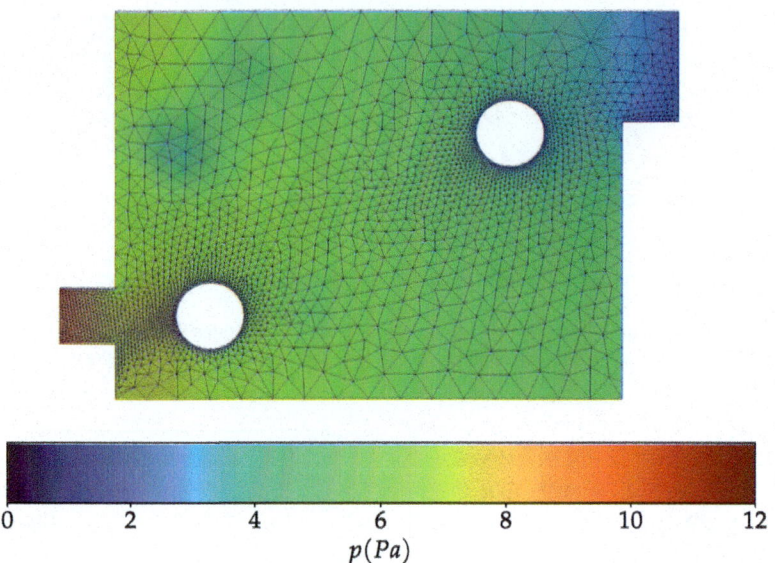

Fig. 13.7 Pressure field as computed by the above program Sect. 13.5.4 at the final time step, $t = 3.0s$. The gradient of the pressure is noticeable from the inlet to the outlet. Two low-pressure regions are also visible toward the top left

named "dset" with a dictionary structure and a key name "u" is created. The shape of this dataset is a vector of length "usize".

In lines 282–285, a progress bar using the "tqdm" library is specified.

Line 291 marks the start of the time loop. In line 304 we specify the update interval of the progress bar. In line 307 we increment the initial time with Δt, then in line 310–311, we evaluate a function value for "u_at_inlet" for the incremented time.

The first weak form is solved for "u_s" in lines 313–330. From lines 332–346, we solve for variable "p_" and in lines 348–355, we solve for "u_" again. This is the correction step. The solved function values are written to the hard disk in lines 358 and 359. In lines 360 and 361, we store the corrected velocity data as an entry in the "dset" dictionary. This velocity data is 1500 in number corresponding to the total number of time steps.

In lines 363–368, we assign the values of "u_" into "u_n" and the values of "u_n" into "u_n1". Finally, we close the file handles in lines 371–373.

From lines 376–394, we check the presence of working directories, files, result folders, and mesh files; they are created as and when required and deemed necessary (Fig. 13.7).

```
1    """
2    Program for Navier-Stokes simulation
3    on a 2D domain with 1 inlet and 1 outlet
4    """
5    import os
6    import sys
7    import ufl
8    from ufl import (div, dot, dx, inner,
9                        grad, nabla_grad,
10                       lhs, rhs)
11   from dolfinx.io.gmshio import read_from_msh
12   from dolfinx.io import XDMFFile
13   from dolfinx import fem
14   from petsc4py import PETSc
15   from mpi4py import MPI
16   import tqdm.autonotebook
17   import numpy as np
18   import h5py
19
20   """
21   File handling block: We load an externally generated
22   mesh from GMSH. We also specify the location of the
23   mesh (.msh) file and the location where the results
24   would be stored.
25   """
26   simulationName = "Stream_NS-2D"
27   meshName = "aquifer2D"
28   meshPath = "./meshes_gmsh/" + meshName + ".msh"
29   resultPath = "./result/" + simulationName + "/"
30
31   if os.path.isfile(meshPath):
32       """
33       Load external mesh that was generated using GMSH.
34       The external mesh has dimension=3 although the
35       z-coordinates are just zeros. Therefore, we use
36       gdim=2, to specify the actual dimension of the
37       problem.
```

```python
"""
gdim = 2
domain, cell_tags, facet_tags = read_from_msh(
    filename=meshPath,
    comm=MPI.COMM_WORLD,
    rank=0, gdim=gdim)

    # Integer markers are set during mesh creation in GMSH
inlet_marker = 13
outlet_marker = 14
noflow_marker = 15
island_marker = 16

# Facet dimension for boundary condition
fdim = domain.topology.dim - 1


# ------------------------------------------------------------
"""
Setting material properties and simulation times.
Kinematic viscosity and density are the only
parameters that are specified.
"""
# Representative Water Kinematic viscosity
mu_water_rep = fem.Constant(domain=domain,
                            c=PETSc.ScalarType(0.0089))
# Representative Water Density
rho_water_rep = fem.Constant(domain=domain,
                             c=PETSc.ScalarType(1.0))
t_initial = 0  # Time initial
t_final = 3  # Time final
dt = 1 / 1500  # Stepping
n_steps = int((t_final - t_initial) / dt)  # N steps
k = fem.Constant(domain=domain,  # Convert to PETSc type
                 c=PETSc.ScalarType(dt))


"""
Finite elements and Function spaces: We specify
two element types; vector Lagrange element for the
velocity field and a simple Lagrange element for
the pressure field.
"""
LG2_elem = ufl.VectorElement(family="Lagrange",
                             cell=domain.ufl_cell(),
                             degree=2)
LG1_elem = ufl.FiniteElement(family="Lagrange",
                             cell=domain.ufl_cell(),
                             degree=1)
Vspace = fem.FunctionSpace(mesh=domain, element=LG2_elem)
```

```python
86      Qspace = fem.FunctionSpace(mesh=domain, element=LG1_elem)
87
88
89      class VelocityProfileAtInlet:
90          """
91          Time dependent Boundary Condition: We intend to
92          vary the inlet velocity to have a profile
93          according to a given equation. This is realized
94          using a time-dependent boundary condition helping
95          to change the velocity as a function of time.
96          """
97
98          def __init__(self, t):
99              self.t = t
100
101          def __call__(self, x):
102              velocity_vector = np.zeros((2, x.shape[1]),
103                                          dtype=PETSc.ScalarType)
104              velocity_vector[0] = 10.0 * (1.001 + np.sin(
105                  2 * self.t * np.pi / 3
106              )) * x[1] * (3.5 - x[1]) / (3.5 ** 2)
107              return velocity_vector
108
109
110  # BC at Inlet
111  u_at_inlet = fem.Function(V=Vspace)
112  inlet_velocity = VelocityProfileAtInlet(t=t_initial)
113  u_at_inlet.interpolate(u=inlet_velocity)
114  bc_u_inflow = fem.dirichletbc(
115      value=u_at_inlet,
116      dofs=fem.locate_dofs_topological(
117          V=Vspace,
118          entity_dim=fdim,
119          entities=facet_tags.find(inlet_marker)
120      )
```

```
121   )
122
123   # No-slip (at Walls)
124   u_noslip = np.array(
125       tuple([0, ]) * gdim, dtype=PETSc.ScalarType)
126   bc_u_walls = fem.dirichletbc(
127       value=u_noslip,
128       dofs=fem.locate_dofs_topological(
129           V=Vspace, entity_dim=fdim,
130           entities=facet_tags.find(noflow_marker)
131       ),
132       V=Vspace
133   )
134
135   # No-slip (on Islands)
136   bc_u_island = fem.dirichletbc(
137       value=u_noslip,  # as this vector is already created
138       dofs=fem.locate_dofs_topological(
139           V=Vspace,
140           entity_dim=fdim,
141           entities=facet_tags.find(island_marker)
142       ),
143       V=Vspace)
144   bc_on_velocity = [bc_u_inflow,
145                     bc_u_island,
146                     bc_u_walls]
147
148   # BC at Outlet
149   bc_p_outlet = fem.dirichletbc(
150       value=PETSc.ScalarType(0.0),
151       dofs=fem.locate_dofs_topological(
152           V=Qspace,
153           entity_dim=fdim,
154           entities=facet_tags.find(outlet_marker)
155       ),
156       V=Qspace)
157   bc_on_pressure = [bc_p_outlet]
158
159   """
160   Defining Test and Trial Functions: We define
161   the trial and test functions to be used in the
162   weak form.
163   """
```

```
164    u = ufl.TrialFunction(function_space=Vspace)
165    v = ufl.TestFunction(function_space=Vspace)
166
167    u_ = fem.Function(V=Vspace)
168    u_.name = "velocity"
169
170    u_s = fem.Function(V=Vspace)
171
172    u_n = fem.Function(V=Vspace)
173    u_n1 = fem.Function(V=Vspace)
174
175    p = ufl.TrialFunction(function_space=Qspace)
176    q = ufl.TestFunction(function_space=Qspace)
177    p_ = fem.Function(V=Qspace)
178    p_.name = "pressure"
179
180    phi = fem.Function(V=Qspace)
181
182    # To be used in the weak-form
183    f = fem.Constant(
184        domain=domain,
185        c=PETSc.ScalarType((0.0, 0.0)))
186
187    """
188    Specifying the variational formulation
189    """
190    F = rho_water_rep / k * dot(
191        u - u_n, v) * dx
192    F = F + inner(
193        dot(
194            (3 / 2) * u_n - (1 / 2) * u_n1,
195            (1 / 2) * nabla_grad(u + u_n)
196        ), v
197    ) * dx
198    F = F + (1 / 2) * mu_water_rep * inner(
199        grad(u + u_n), grad(v)) * dx
200    F = F - dot(
201        p_, div(v)) * dx
202    F = F + dot(
```

```
203          f, v) * dx
204
205     # Extracting the LHS and RHS for the linear
206     # system
207     a1 = fem.form(lhs(F))
208     L1 = fem.form(rhs(F))
209
210     a2 = fem.form(dot(grad(p),
211                         grad(q)) * dx)
212     L2 = fem.form(-rho_water_rep / k * dot(div(u_s),
213                                             q) * dx)
214
215     a3 = fem.form(rho_water_rep * dot(u, v) * dx)
216     L3 = fem.form(rho_water_rep * dot(
217         u_s, v) * dx - k * dot(nabla_grad(phi),
218                                 v) * dx)
219
220     """
221     Forming linear system
222     """
223     A1 = fem.petsc.create_matrix(a=a1)
224     b1 = fem.petsc.create_vector(L=L1)
225
226     A2 = fem.petsc.assemble_matrix(
227         a2, bc_on_pressure)
228     b2 = fem.petsc.create_vector(L=L2)
229
230     A3 = fem.petsc.assemble_matrix(a3)
231     b3 = fem.petsc.create_vector(L=L3)
232
233     A2.assemble()
234     A3.assemble()
235
236     """
237     Configuring solvers for Steps 1, 2 and 3
238     """
239     STEP1_solver = PETSc.KSP().create(domain.comm)
240     STEP1_solver.setOperators(A1)
241     STEP1_solver.setType(PETSc.KSP.Type.BCGS)
242     STEP1_solver.getPC().setType(PETSc.PC.Type.JACOBI)
```

```
243    STEP2_solver = PETSc.KSP().create(domain.comm)
244    STEP2_solver.setOperators(A2)
245    STEP2_solver.setType(PETSc.KSP.Type.MINRES)
246    STEP2_solver.getPC().setHYPREType("boomeramg")
247    STEP2_solver.getPC().setType(PETSc.PC.Type.HYPRE)
248
249    STEP3_solver = PETSc.KSP().create(domain.comm)
250    STEP3_solver.setOperators(A3)
251    STEP3_solver.setType(PETSc.KSP.Type.CG)
252    STEP3_solver.getPC().setType(PETSc.PC.Type.SOR)
253
254    """
255    Creating file objects for storing results
256    """
257    xdmfu = XDMFFile(comm=domain.comm,  # velocity result
258                     filename=resultPath + "stream_u.xdmf",
259                     file_mode="w",
260                     encoding=XDMFFile.Encoding.HDF5)
261    xdmfu.write_mesh(mesh=domain)
262    xdmfu.write_function(u=u_,
263                         t=t_initial)
264
265    xdmfp = XDMFFile(comm=domain.comm,  # pressure result
266                     filename=resultPath + "stream_p.xdmf",
267                     file_mode="w",
268                     encoding=XDMFFile.Encoding.HDF5)
269    xdmfp.write_mesh(mesh=domain)
270    xdmfp.write_function(u=p_,
271                         t=t_initial)
272
273    # for storing the time series velocity data
274    usize = u_.x.array.shape[0]
275    with h5py.File(
276            name=resultPath + 'velocity_timeseries.h5',
277            mode='w') as fr:
278        dset = fr.create_dataset(name="u",
279                                 shape=[n_steps, usize],
280                                 dtype=np.float32)
281        progress = tqdm.autonotebook.tqdm(
282            desc="Solving PDE system",
283            total=n_steps
284        )
285
286        """
287        Time stepping through solutions
288        """
```

```
289      t_at = 0
290      for i in range(n_steps):
291          """
292          Following operations are done:
293          1) Update the new time
294          2) Get the new velocity at the source
295          3) Compute the tentative velocity
296          4) Do the pressure correction
297          5) Do the velocity correction
298          6) Save the solutions
299          7) Store the result for next time step
300          """
301
302          # 0) Tracking progress at every 10 steps
303          progress.update(n=1)
304
305          # 1) Update current time step
306          t_at = t_at + dt
307
308          # 2) Update inlet velocity
309          inlet_velocity.t = t_at
310          u_at_inlet.interpolate(u=inlet_velocity)
311
312          # 3) Estimating velocity
313          A1.zeroEntries()
314          fem.petsc.assemble_matrix(
315              A1, a1, bcs=bc_on_velocity)
316          A1.assemble()
317
318          with b1.localForm() as loc:
319              loc.set(0)
320          fem.petsc.assemble_vector(b1, L1)
321          fem.petsc.apply_lifting(
322              b=b1, a=[a1], bcs=[bc_on_velocity])
323          b1.ghostUpdate(
324              addv=PETSc.InsertMode.ADD_VALUES,
325              mode=PETSc.ScatterMode.REVERSE)
326          fem.petsc.set_bc(
327              b=b1, bcs=bc_on_velocity)
328          STEP1_solver.solve(b1, u_s.vector)
329          u_s.x.scatter_forward()
330
331          # 4): Doing Pressure correction
```

```
332        with b2.localForm() as loc:
333            loc.set(0)
334        fem.petsc.assemble_vector(b2, L2)
335        fem.petsc.apply_lifting(
336            b=b2, a=[a2], bcs=[bc_on_pressure])
337        b2.ghostUpdate(addv=PETSc.InsertMode.ADD_VALUES,
338                       mode=PETSc.ScatterMode.REVERSE)
339        fem.petsc.set_bc(
340            b=b2, bcs=bc_on_pressure)
341        STEP2_solver.solve(b2, phi.vector)
342        phi.x.scatter_forward()
343
344        p_.vector.axpy(1, phi.vector)
345        p_.x.scatter_forward()
346
347        # 5) Doing Velocity correction
348        with b3.localForm() as loc:
349            loc.set(0)
350        fem.petsc.assemble_vector(b3, L3)
351        b3.ghostUpdate(addv=PETSc.InsertMode.ADD_VALUES,
352                       mode=PETSc.ScatterMode.REVERSE)
353        STEP3_solver.solve(b3, u_.vector)
354        u_.x.scatter_forward()
355
356        # 6) Storing the solutions
357        xdmfu.write_function(u_, t_at)
358        xdmfp.write_function(p_, t_at)
359        u_data = u_.x.array
360        dset[i] = u_data
361
362        # 7) Assign future with current state
363        with u_.vector.localForm() as loc_, \
364                u_n.vector.localForm() as loc_n, \
365                u_n1.vector.localForm() as loc_n1:
366            loc_n.copy(loc_n1)
367            loc_.copy(loc_n)
368
369    # Close the file handles
370    xdmfu.close()
371    xdmfp.close()
372 fr.close()
```

```
373      print("Done!")

374

375   else:
376        """
377        1) Check if (.msh) file exists or not?
378        2) Recommend to put GMSH file in meshPath.
379        """
380        if not os.path.exists(meshPath):
381            print("\n\nMsg: Source path created!")
382            os.makedirs(meshPath)

383

384        if not os.path.exists(resultPath + "/"):
385            """
386            1) Create result folder if required.
387            """
388            os.makedirs(resultPath + "/")
389            print("Msg: Result folder created!")

390

391        if not os.path.isfile(meshPath + meshName + ".msh"):
392            print("Msg: Check if GMSH mesh file (*.msh) exists?")
393            sys.exit()
```

13.5.5   Post-processing of the Results

All the post-processing has been done in Paraview software. The program in Sect.
13.5.4 saves the results in ".xdmf" format. These files are open-source file formats
that can be readily loaded by Paraview. The file contains simulated data for each
time step from $t = 0.00$ to $t = 3.00$ in 1500 time steps. This has been kept the
same as the Navier-Stokes simulation. Figure 13.4 has been plotted by selecting
the "Surface" type representation and "Coloring" by the variable's magnitude in the
loaded mesh. The Fig. 13.4a–f were generated by selecting appropriate time values
from the "Time" dropdown menu in the software. The Paraview application uses
"Cool to Warm" colormap by default. A better alternative is the "Turbo" colormap
which has been used to draw the figures here. One can see the spreading of the
contaminant even at end-time values as a result of using this versatile colormap.

13.5.6   Discussion of the Results

In this section, we have done a 2D Navier-Stokes simulation for an incompressible
fluid flow. The strong form of the equations is given in Sect. 13.5. The variational
form of the equation is given in Sect. 13.5.3. It consists of three linear problems
that are solved sequentially; the result is that we get the pressure and velocity field
variables for every time step in the time loop.

Velocity field:

In Fig. 13.4a–f, we see the velocity fields that correspond to time steps, $t = 0.03s, 0.33s, 1.0s, 1.66s, 2.33s, 3.00s$. At $t = 0.03s$, in sub-figure (a), the velocity at the inlet appears to be higher than the outlet; the color intensity stands for the said comparison. We can also see a bifurcation in the flow when it hits the obstacle at the bottom-left. There is a similar structure around the obstacle near the outlet at the top-right. At the next time step, at $t = 0.33s$, the flow advances some more and the bifurcated flow structure becomes prominent; high-intensity (dark red) regions are found to cover more area. At the same time, we can see the beginning of the vortex street near the left bank of the domain.

At the time step, $t = 1.0s$, the high-intensity regions cover still more space in the domain. The vortex becomes more evident toward the left. The bifurcation in the flow is more pronounced; one can see elongated flow streams. The velocity intensities remain more or less the same at the outlet.

At time step, $t = 1.66s$, the flow stream starts to convert to a fully developed vortex toward the left-top. The flow around the second obstacle appears to exhibit a triangle-shaped low-velocity zone in the northeast direction toward the outlet direction. Even the center appears to have a low-velocity zone.

At time step, $t = 2.33s$, we can see a fully developed vortex at both the top and bottom of the obstacle present at the left bottom. A low-velocity zone (almost zero magnitude) is found to cover the right side of the domain.

At time step, $t = 3.00s$, the vortex in the left-top exhibits a spiral shape, and the low-velocity region at the right gets modified. A small magnitude velocity stream is present at the outlet. A few low-velocity spots can also be seen spread across the domain.

Statistical Analysis:

Once the simulations are complete we intend to perform a statistical analysis of the velocity data collected from the inlet and the outlet. For this, we select a group of nodes at the inlet and the outlet (Fig. 13.5). The pink-colored nodes mark the probes for which the velocity data is acquired for all the 1500 time steps. We compute the minimum, maximum, first quartile, third quartile, and the mean of the data. The result of the analysis has been plotted in Fig. 13.6.

A velocity profile having a sinusoid shape was initiated and run. The resultant profile at the outlet seems to follow some part of the profile at the inlet. The velocity at the outlet also appears to be reduced and is almost half of the input.

Pressure field:

We also show the pressure field at the time step, $t = 3.00s$. One can notice two low-pressure zones at the top-left of the domain. The physical positions of the two zones corroborate with the velocity field as shown in Fig. 13.4f. The inlet and the outlet have high and low-pressure regions, respectively.

Chapter 14
Contaminant Transport Models

In this chapter, we develop three finite element models for common contaminant transport problems. The cases represent different scenarios of contaminant transport. Advection is an important process in contaminant transport problems. For the models with an advection component, we use the velocity field data (saved a priori) of the Navier-Stokes simulation from Sect. 13.5.

14.1 A 2D Diffusion Reaction Model

In this section, we develop a 2D diffusion, reaction model. Specifically, the code presented solves a scenario where a contaminant is undergoing diffusion as well as reacting with the environment with time. The result is the spreading of the contaminant from the location of the source function.

14.1.1 Variational Formulation

The strong form of the diffusion-reaction model can be specified using the following governing equation:

$$\frac{\partial u}{\partial t} + \nabla \cdot (D\nabla_u) = f - Ru \tag{14.1}$$

where u is the concentration of the contaminant, the constant D is the diffusion coefficient, and R is the rate of reaction. The weak form corresponding to Eq. 14.1 is given as

A. Kumar and M. Saharia, *Python for Water and Environment*, Innovations in Sustainable Technologies and Computing, https://doi.org/10.1007/978-981-99-9408-3_14

$$\int_\Omega \frac{1}{dt} \langle u - u_n, v \rangle dx + \int_\Omega D \langle \nabla u, \nabla v \rangle dx = \int_\Omega (fv - Ruv) dx \qquad (14.2)$$

where u, u_n are the values of the concentration at the current and previous time steps; v is the test function defined in the function space of the concentration variable; and f denotes the source function.

14.1.2 Python Code

In this section, we define the code for diffusion, advection processes for a contaminant transport problem. We begin by importing the necessary libraries in lines 8–16.

In lines 19–25, the name of the simulation, the location of the externally generated mesh file, and the path to save the results are specified.

In lines 27–40, the operation of loading the mesh file with the ".msh" extension is performed. Along with the mesh, we also read the cell and facet tags. The dimension of the domain is 2, therefore the facet dimension is 1 less than the domain.

In lines 42–51, we define the parameters for the time loop such as the start, stop, Δt, and the number of steps.

In lines 53–65, we define the constants that are to be used in the weak formulation; they are Δt, the rate of the reaction, and the diffusion coefficient.

In lines 67–84, we define the function spaces, finite element type, and the functions that are later used in the weak formulation. The type of the element is "Lagrange" and the name of the function space is "V". We further define a test function named "v1" and two functions—"u1" and "u1n"—for working with the concentration variable (Fig. 14.1).

```
1    """
2    Program to simulate the diffusion reaction
3    system in the domain realized using the
4    aquifer geometry of the Navier-Stokes simulation.
5    """
6
7
8    # We first load the necessary libraries
9    import numpy as np
10   from mpi4py import MPI
11   from petsc4py import PETSc
12   import ufl
13   from dolfinx import fem, nls, io
14   from ufl import (dot, dx, grad)
15   from dolfinx.io import XDMFFile
```

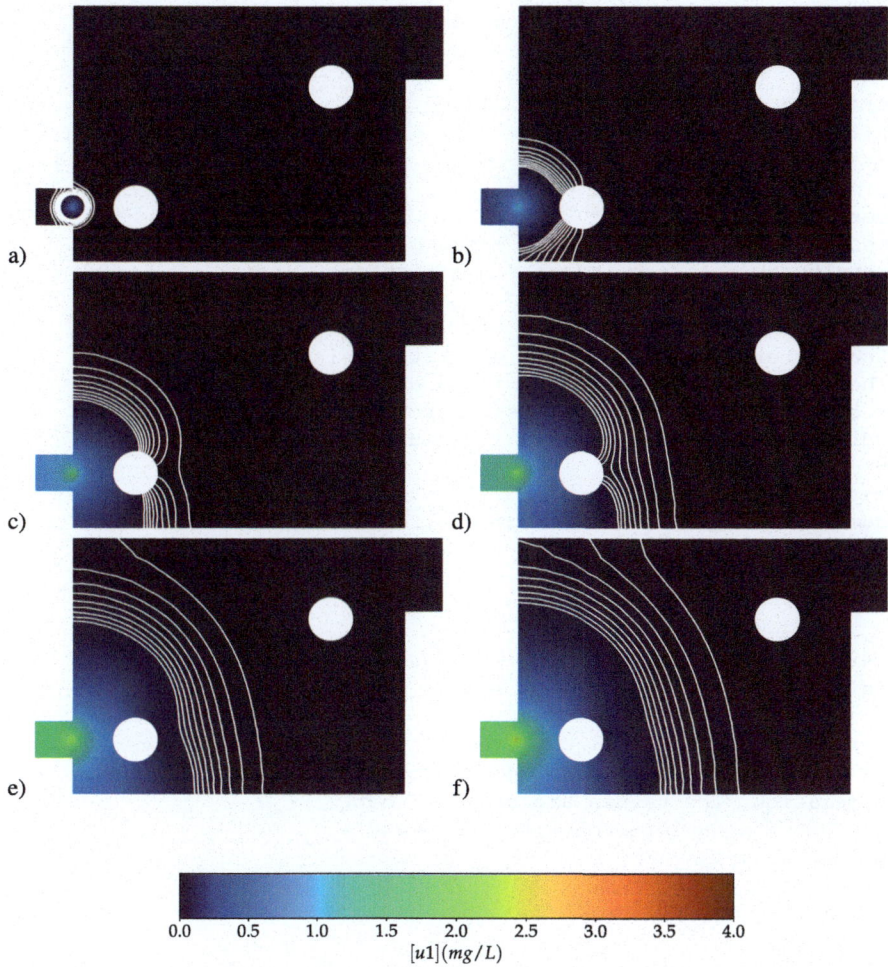

Fig. 14.1 Concentration field as computed by Program Sect. 14.1.2 at 6 different times, $t = 0.03$ s, 0.33 s, 1.0 s, 1.66 s, 2.33 s, 3.00 s. One can notice specific patterns that emerge due to the presence of the two circular-shaped islands imposing the no-slip boundary condition along with the surrounding boundaries

```python
16   from dolfinx.io.gmshio import read_from_msh
17
18
19   """
20   File handling
21   """
22   simulationName = "Diff_React-2D"
23   meshName = "aquifer2D"
24   meshPath = "./meshes_gmsh/" + meshName + ".msh"
25   resultPath = "./result/" + simulationName + "/"
26
27   """
28   Load external mesh that was generated using GMSH.
29   The external mesh has dimension=3 although the
30   z-coordinates are just zeros. Therefore, we use
31   gdim=2, to specify the actual dimension of the
32   problem.
33   """
34   domain, cell_tags, ft = io.gmshio.read_from_msh(
35       filename=meshPath,
36       comm=MPI.COMM_WORLD,
37       rank=0,
38       gdim=2)
39   gdim = domain.topology.dim   # domain dimension
40   fdim = gdim - 1   # facet dimension
41
42   """
43   We define the start and stop time of the
44   simulation. The values are same as that
45   used in the Navier-Stokes to help make a
46   comparison.
47   """
48   t_start = 0.0
49   t_stop = 3.0
50   delta_t = 1 / 1500
51   n_steps = int((t_stop - t_start) / delta_t)   # N steps
52
53   """
54   These constants are defined as they are
55   to be used in the weak form of the PDE.
56   """
57   dt_inv = fem.Constant(
58       domain=domain,
59       c=PETSc.ScalarType(1 / delta_t))
60   Rate = fem.Constant(
61       domain=domain,
62       c=PETSc.ScalarType(0.01))
63   D_coeff = fem.Constant(
64       domain=domain,
```

```
65        c=PETSc.ScalarType(0.3))
66
67    """
68    Here, we define the element type
69    and the function space to solve
70    for the concentration of the
71    reactant.
72    """
73    P_elem = ufl.FiniteElement(
74        family="Lagrange",
75        cell=domain.ufl_cell(),
76        degree=1)
77    V = fem.FunctionSpace(
78        mesh=domain,
79        element=P_elem)
80    v1 = ufl.TestFunction(function_space=V)
81    u1 = fem.Function(V=V)
82    u1n = fem.Function(V=V)
83    u1.name = "concentration"
84    u1n.name = "concentration"
85
86
87    class SourceExpression:
88        """
89        We define the source location and its
90        magnitude that would go into the weak
91        form. The (x, y) location is specified
92        using 'ptSrc_xy' argument.
93        The magnitude is specified as 10.0.
94        """
95
96        def __init__(self, t_, ptSrc_xy):
97            self.t = t_
98            self.ptSrc_xy = ptSrc_xy
99
100       def eval(self, x):
101           values = np.full(x.shape[1], 0.0)
102           idx = ((x[0] - self.ptSrc_xy[0]) ** 2) + \
103                 ((x[1] - self.ptSrc_xy[1]) ** 2) < 0.05 ** 2
104           values[idx] = 100.0  # Strength of the source
105           return values
106
107
108   # Inheriting from the source function class definition
```

```
109   source = SourceExpression(t_=0.0,
110                             ptSrc_xy=[0.50, 0.75])
111
112   # Interpolating into a function
113   f = fem.Function(V=V)
114   f.interpolate(u=source.eval)
115
116   # Variational formulation
117   F = ((u1 - u1n) * dt_inv) * v1 * dx
118   F = F + D_coeff * dot(grad(u1), grad(v1)) * dx
119   F = F + Rate * u1 * v1 * dx
120   F = F - f * v1 * dx
121
122   """
123   The current diffusion reaction equation,
124   although solvable by a linear solver, we
125   deliberately do away with a nonlinear
126   solver because one may wish to model a
127   case where the rate of decomposition of
128   the contaminant may at times be nonlinear.
129   Newton solver is good in such cases.
130   """
131   problem = fem.petsc.NonlinearProblem(
132       F=F,
133       u=u1)
134   solver = nls.petsc.NewtonSolver(
135       comm=MPI.COMM_WORLD,
136       problem=problem)
137   solver.rtol = 1e-6
138   solver.report = True
139
140   """
141   We use the open file format '.xdmf' to store
142   the results of the simulation. We first write
143   the mesh followed by the function. The mesh is
144   written just once; then in the time loop we
145   only write the function values as the mesh can
146   now be shared during visualization.
147   """
```

```
148  xdmfu = XDMFFile(
149      comm=domain.comm,
150      filename=resultPath + simulationName + ".xdmf",
151      file_mode="w",
152      encoding=XDMFFile.Encoding.HDF5)
153  xdmfu.write_mesh(mesh=domain)
154  xdmfu.write_function(
155      u=u1,
156      t=t_start)
157
158  """
159  There is no velocity term in this system. We only
160  model the contaminant spread due to diffusion.
161  """
162
163  t = t_start
164  for n in range(n_steps):
165      print("Doing {}/{} step".format(n, n_steps))
166      t = t + delta_t
167      r = solver.solve(u=u1)
168      u1n.x.array[:] = u1.x.array
169      xdmfu.write_function(
170          u=u1,
171          t=t)
172  xdmfu.close()
173  print("Done!")
```

In lines 87–105, the class definition of the point source is expressed. The "__init__" method is used to initialize an instance of the class. The "__eval__" method is used to evaluate the function at designated indices. A source with a radius of 0.05 is specified, and the function evaluates to 100 for all coordinates that lie within this radius. The class can be invoked with the location of the point source and the time. In line 104, the strength of the contaminant source has been specified (currently 100.0); the class is invoked in lines 109 and 110. A function, "f" which is defined on the function space "V" is interpolated with the source expression.

In lines 116–120, we define the weak formulation of the problem. There is no advection term in the formulation. The diffusion coefficient, rate of the reaction, and the strength of the source terms play a significant role in the simulation.

In lines 122–138, we specify the type of problem as nonlinear with the solver type as "Newton".

In lines 141–156, the file operation object is specified. We first write the mesh and the associated time. The file mode is mentioned as "w" for writing operation and the encoding is "hierarchical data format or HDF5".

In lines 158–173, we specify the time loop that runs from "n = 0" to "n = 1499". At each iteration we first increment the current time "t" with "Δt"; solve for variable "u1"; and assign the solution to "u1n" again with the associated time step. At the next iteration the stored value in "u1n" is used as the previous solution.

14.1.3   Post-processing

All the post-processing has been done in the Paraview software. The program in Sect. 14.1.2 saves the results in ".xdmf" format. These files are open-source file formats that can be readily loaded by Paraview. The file contains simulated data for each time step from $t = 0.00$ to $t = 3.00$ in 1500 time steps. This has been kept the same as the Navier-Stokes simulation. Figure 14.1 has been plotted by selecting the "Surface"-type representation and "Coloring" by the variable's magnitude in the loaded mesh. Figure 14.1a–f was generated by selecting appropriate time values from the "Time" dropdown menu in the software. The Paraview application uses "Cool to Warm" colormap by default. A better alternative is the "Turbo" colormap which has been used to draw the figures here. One can see the spreading of the contaminant even at end-time values as a result of using this versatile colormap.

14.1.4   Discussion of the Results

Figure 14.1 shows the result of simulating a contaminant transport problem governed by a diffusion and reaction process. The contours in the figures show the periphery of the spread of the contaminant. One can observe that, with time, the concentration value gradually decreases as one moves away from the source which is specified at $(x, y) = (0.50, 0.75)$. The spread appears to be uniformly distributed at all time steps; this is due to the absence of any advection term in the governing equations. The presence of any advection term would have disrupted the uniformity. As viewed later in the sections we can see the modified patterns when an advection term is included.

14.2   A 2D Diffusion Advection Model

In this section, we model a scenario where a single contaminant is undergoing diffusion as well as advection. Such models are well suited where the movement of water is involved in the transportation of the contaminant. This movement can be readily associated with a velocity in the X- and Y-directions which consequently show in an advection process.

14.2.1   *Variational Formulation*

The strong form of the governing equation of diffusion, advection scenario can be given as

$$\frac{\partial [u]}{\partial t} + \mathbf{w} \cdot \nabla[u] + \nabla \cdot (D\nabla[u]) = f \qquad (14.3)$$

where $[u]$ is the concentration of the contaminant, $\mathbf{w}$ is the velocity of the flow, and constant D is the diffusion coefficient. The weak form corresponding to Eq. 14.3 can be stated as

$$\int_{\Omega} \frac{1}{dt} \langle u - u_n, v \rangle dx + \int_{\Omega} \langle \mathbf{w} \cdot \nabla u, v \rangle dx + \int_{\Omega} D \langle \nabla u, \nabla v \rangle dx$$
$$= \int_{\Omega} (fv) dx \qquad (14.4)$$

14.2.2   *Python Code*

The code for an advection, diffusion process is listed in this section. We begin by loading the essential libraries in lines 13–23. From lines 25–33, we set the names for the simulation type and specify a location for the externally generated mesh and a path for saving the results.

In lines 36–47, the mesh is loaded along with the cell tags and facet tags; variables "gdim" and "fdim" correspond to the dimension of the domain and the facet.

In lines 50–58, simulation start and stop times have been provided. The number of steps in between has been given as 1500.

From lines 60 to 69, we define the constants for the inverse of change in time and diffusion coefficient with their types as "PETSc scalar". This is in favor of parallel processing; in the current programs, all operations are serialized.

In lines 71–88, we define the finite element type as "Lagrange" and the cell type has been inferred from the imported mesh; the cell type is triangular. The concentration variable is a scalar and hence the finite element type. We have also defined a test function called "v1" on the function space "V". "u1" and "u1n" are functions to hold the values of the solution of the solver. We do away with the trial functions as we treat the problem as nonlinear implying the usage of Newton solver of "dolfinx".

```python
1    """
2    Program to solve diffusion, advection
3    equation for modeling the contaminant
4    transport problem. Given the location
5    and strength (concentration) of the
6    source function, we study the concentration
7    of the reactant as a function of space and
8    time. To make the visualizations appealing
9    we load the velocity data from the
10   Navier-Stokes simulation output.
11   """
12
13   # We first load the necessary libraries
14   import h5py
15   import numpy as np
16   from mpi4py import MPI
17   from petsc4py import PETSc
18   import ufl
19   from dolfinx import fem, nls, io
20   from ufl import (dot, dx, grad)
21   from dolfinx.io import XDMFFile
22   from dolfinx.io.gmshio import read_from_msh
23   import tqdm.autonotebook
24
25   """
26   File handling block
27   """
28   simulationName = "Diff_Adv-2D"
29   meshName = "aquifer2D"
30   meshPath = "./meshes_gmsh/" + meshName + ".msh"
31   velocityDataPath = "./result/" \
32                       "Stream_NS-2D/velocity_timeseries.h5"
33   resultPath = "./result/" + simulationName + "/"
34
35
36   """
37   We load the same mesh, generated externally
38   using GMSH which was also used for the
39   Navier-Stokes simulation.
40   """
41   domain, cell_tags, ft = io.gmshio.read_from_msh(
42       filename=meshPath,
43       comm=MPI.COMM_WORLD,
44       rank=0,
45       gdim=2)
46   gdim = domain.topology.dim  # domain dimension
47   fdim = gdim - 1  # facet dimension
```

```
48    """
49
50    We define the start and stop time of the
51    simulation. The values are same as that
52    used in the Navier-Stokes because we want
53    to read the velocity data it.
54    """
55    t_start = 0.0
56    t_stop = 3.0
57    delta_t = 1 / 1500
58    n_steps = int((t_stop - t_start) / delta_t)   # N steps
59
60    """
61    These constants are defined as they are
62    to be used in the weak form of the PDE.
63    """
64    dt_inv = fem.Constant(
65        domain=domain,
66        c=PETSc.ScalarType(1 / delta_t))
67    D_coeff = fem.Constant(
68        domain=domain,
69        c=PETSc.ScalarType(0.03))
70
71    """
72    Here, we define the element type
73    and the function space to solve
74    for the concentration of the
75    reactant.
76    """
77    P1 = ufl.FiniteElement(
78        family="Lagrange",
79        cell=domain.ufl_cell(),
80        degree=1)
81    V = fem.FunctionSpace(
82        mesh=domain,
83        element=P1)
84    v1 = ufl.TestFunction(function_space=V)
85    u1 = fem.Function(V=V)
86    u1n = fem.Function(V=V)
87    u1.name = "concentration"
```

```python
88    u1n.name = "concentration"
89
90    """
91    Here, we define a vector element
92    and a corresponding function space,
93    function for loading the
94    velocity (a vector) data from the
95    NS simulation.
96    """
97    Pvec = ufl.VectorElement(
98        family="Lagrange",
99        cell=domain.ufl_cell(),
100       degree=2)
101   W = fem.FunctionSpace(
102       mesh=domain,
103       element=Pvec)
104       w = fem.Function(V=W)
105
106
107   class SourceExpression:
108       """
109       We define the source location and its
110       magnitude that would go into the weak
111       form. The (x, y) location is specified
112       using 'ptSrc_xy' argument.
113       The magnitude is specified as 10.0.
114       """
115       def __init__(self, t_, ptSrc_xy):
116           self.t = t_
117           self.ptSrc_xy = ptSrc_xy
118
119       def eval(self, x):
120           values = np.full(x.shape[1], 0.0)
121           idx = ((x[0] - self.ptSrc_xy[0]) ** 2) + \
122                 ((x[1] - self.ptSrc_xy[1]) ** 2) < 0.05 ** 2
123           values[idx] = 100.0  # Strength of the source
124           return values
125
126
127   # Inheriting from the source function class definition
128   source = SourceExpression(t_=0.0,
129                             ptSrc_xy=[0.25, 0.75])
130
131   # Interpolating into a function
132   f = fem.Function(V=V)
```

```
133    f.interpolate(u=source.eval)
134
135    """
136    Here, we define the weak formulation
137    of the partial differential equation.
138    'w' is a function defined on a vector
139    function space 'W' that holds the value
140    of the velocity field. This velocity
141    is taken from a previous Navier-Stokes
142    simulation.
143    """
144    F = ((u1 - u1n) * dt_inv) * v1 * dx
145    F = F + dot(w, grad(u1)) * v1 * dx
146    F = F + D_coeff * dot(grad(u1), grad(v1)) * dx
147    F = F - f * v1 * dx
148
149    """
150    The current advection reaction equation,
151    although solvable by a linear solver, we
152    deliberately do away with a nonlinear
153    solver because one may wish to model a
154    case where the rate of decomposition of
155    the contaminant may at times be nonlinear.
156    Newton solver is good with them.
157    """
158    problem = fem.petsc.NonlinearProblem(
159        F=F,
160        u=u1)
161    solver = nls.petsc.NewtonSolver(
162        comm=MPI.COMM_WORLD,
163        problem=problem)
164    solver.rtol = 1e-6
165    solver.report = True
166
167    """
168    We use the open file format '.xdmf' to store
169    the results of the simulation. We first write
170    the mesh followed by the function. The mesh is
171    written just once; in the time loop we only
172    write the function values as the mesh gets
173    shared during visualization.
174    """
```

```
175   xdmfu = XDMFFile(
176       comm=domain.comm,
177       filename=resultPath + simulationName + ".xdmf",
178       file_mode="w",
179       encoding=XDMFFile.Encoding.HDF5)
180   xdmfu.write_mesh(mesh=domain)
181   xdmfu.write_function(
182       u=u1,
183       t=t_start)
184
185   """
186   We load the velocity data file from the Navier-Stokes
187   simulation. The velocity vector (u_i, u_j) is used
188   to realize the advection part of the equation.
189   Because the data was stored in HDF5 format, we employ
190   h5py library to load the data. This is done just before
191   'solving' the equation.
192   """
193   with h5py.File(
194           name=velocityDataPath, mode='r') as fr:
195       print("Name of the velocity variable is {}".format(
196           list(fr.keys())))
197
198       # Time stepping and solving
199       progress = tqdm.autonotebook.tqdm(
200           desc="Solving nonlinear system",
201           total=n_steps
202       )
203       t = t_start
204       for n in range(n_steps):
205           print("Doing {}/{} step".format(n, n_steps))
206           t = t + delta_t
207           u_data = fr['u'][n]
208           w.vector.array = u_data
209           r = solver.solve(u=u1)
210           u1n.x.array[:] = u1.x.array
211           xdmfu.write_function(
212               u=u1,
213               t=t)
214       xdmfu.close()
215   fr.close()
216   print("Done!")
```

From lines 90–104, the codes for the creation of a vector finite element space for loading the pre-saved velocity data of the Navier-Stokes simulation are defined.

In lines 107–124, the class for the source expression is defined via which we define three parameters—the strength of the contaminant source, the location of the source within the domain, and the associated time.

In lines 127–133, the source's class is invoked with an initial value of time, "$t =$ 0.00" and location, $(x, y) = (0.25, 0.75)$. The initial values are used to interpolate the function "f" which is defined on the function space of "V".

From lines 135–147, the weak formulation of the problem is defined. We have ignored any term related to the rate of reaction.

In lines 149–165, we define the problem type as "NonlinearProblem" and specify the solver as "Newton" with a solution tolerance of "1e-6".

In lines 167–183, the objects for a file output operation are defined. ".xdmf" is an open file format supporting speedy I/O operation in parallel; the encoding has been mentioned as "hierarchical data format or HDF5". We first write the mesh and the initial function value of the concentration variable.

From lines 185–216, the time loop after creating the "with" context manager for loading the velocity data using the "h5py" library is specified. For all the 1500 steps and within each iteration of the time loop, the velocity data is first assigned to the "w"; the solver solves for the "u1" and the solution is stored in variable "u1n" to be used in the next iteration.

At last, we close the file handles in lines 214 and 215.

14.2.3   Post-processing

All the post-processing has been done in the Paraview software. The program in Sect. 14.2.2 saves the results in ".xdmf" format. These files are open-source file formats that can be readily loaded by Paraview. The file contains simulated data for each time step from $t = 0.00$ to $t = 3.00$ in 1500 time steps. This has been kept the same as the Navier-Stokes simulation. Figure 14.2 has been plotted by selecting the "Surface"-type representation and "Coloring" by the variable's magnitude in the loaded mesh. Figure 14.2a–f was generated by selecting appropriate time values from the "Time" dropdown menu in the software. The Paraview application uses "Cool to Warm" colormap by default. A better alternative is the "Turbo" colormap which has been used to draw the figures here. One can see the spreading of the contaminant even at end-time values as a result of using this versatile colormap.

14.2.4   Discussion of the Results

In this section, we have demonstrated a diffusion, advection model. Figure 14.2 shows the result of simulating a contaminant transport governed by these two processes. Snapshots of concentration data at a time, $t = 0.03$ s, 0.33 s, 1.0 s, 1.66 s, 2.33 s, 3.00 s, are shown in parts (a) to (f). Due to the advective component of the model, the contaminant is transported into regions of the domain exhibiting specific patterns. It can be seen from the figures that the contaminant gradually covers a larger area within the domain. The source has been specified at a point $(0.25, 0.75)$.

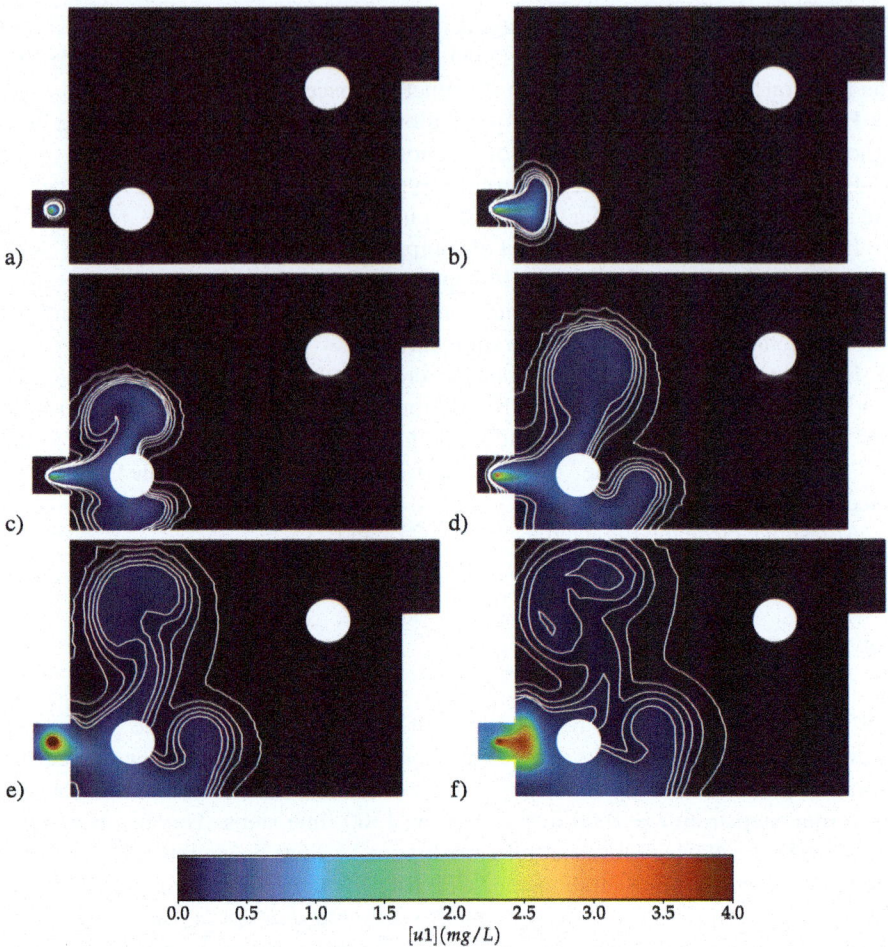

Fig. 14.2 Concentration field as computed by Program Sect. 14.2.2 at 6 different times, $t = 0.03$ s, 0.33 s, 1.0 s, 1.66 s, 2.33 s, 3.00 s. One can notice specific patterns that emerge due to the presence of the two circular-shaped islands imposing the no-slip boundary condition along with the surrounding boundaries

The patterns of transport are due to this; had the source been specified elsewhere we would have expected a different pattern. As this is an advection and diffusion process the parameters—D and $\mathbf{w}$—play a significant role in the simulation.

14.3 Generalized 2D Advection, Diffusion, Reaction Model

In this section, we develop a generalized advection, diffusion, reaction model. We specifically model a scenario where a reactive contaminant is undergoing advection and diffusion while decaying at a rate that is proportional to the concentration, $[u]$ meaning that the rate of reaction follows a first-order kinetics.

14.3.1 Variational Formulation

The strong form of the governing equations for this model is given as

$$\frac{\partial [u]}{\partial t} + \mathbf{w} \cdot \nabla[u] + \nabla \cdot (D\nabla[u]) = f - R[u] \tag{14.5}$$

where $[u]$ is the concentration of the contaminant, w is the velocity of the flow, D is the diffusion coefficient, and R is the rate of reaction.

The weak form corresponding to Eq. 14.5 is given as

$$\int_{\Omega} \frac{1}{dt} \langle u - u_n, v \rangle dx + \int_{\Omega} \langle \mathbf{w} \cdot \nabla u, v \rangle dx + \int_{\Omega} D \langle \nabla u, \nabla v \rangle dx$$
$$= \int_{\Omega} (fv - Ruv) dx \tag{14.6}$$

The variational formulation applies a homogeneous Neumann boundary condition on the entire domain and therefore, in the code, there is no mention of boundary conditions.

14.3.2 Python Code

Here, we give the complete code for advection, diffusion, reaction model. The program is ready to be run after all the necessary libraries have been installed in a designated Conda environment. Although the program has been thoroughly commented on and should be self-explanatory for any modification, we describe some sections that are deemed very important and need explanation.

In lines 13–23, we load the necessary libraries for the program. In lines 26–33, we set the name of the simulation, the path to the external mesh file, the path to the velocity data file, and a path to the results folder where the solution files would be saved. These can be modified by the user to appropriate locations, locally.

```python
"""
Program to solve advection, diffusion, reaction
equations for modeling the contaminant
transport problem. Given the location
and strength (concentration) of the
source function, we study the spread
concentration of the reactant as a function
of space and time. To make the visualizations
appealing we load the velocity data from the
Navier-Stokes simulation output.
"""

# We first load the necessary libraries
import h5py
import numpy as np
from mpi4py import MPI
from petsc4py import PETSc
import ufl
from dolfinx import fem, nls, io
from ufl import (dot, dx, grad)
from dolfinx.io import XDMFFile
from dolfinx.io.gmshio import read_from_msh
import tqdm.autonotebook

"""
File handling block
"""
simulationName = "Adv_Diff_React-2D"
meshName = "aquifer2D"
meshPath = "./meshes_gmsh/" + meshName + ".msh"
velocityDataPath = "./result/" \
                   "Stream_NS-2D/velocity_timeseries.h5"
resultPath = "./result/" + simulationName + "/"

"""
We load the same mesh, generated externally
using GMSH which was also used for the
Navier-Stokes simulation.
"""
domain, cell_tags, ft = io.gmshio.read_from_msh(
  filename=meshPath,
    comm=MPI.COMM_WORLD,
    rank=0,
    gdim=2)
gdim = domain.topology.dim  # domain dimension
```

```python
46   fdim = gdim - 1  # facet dimension
47
48   """
49   We define the start and stop time of the
50   simulation. The values are same as that
51   used in the Navier-Stokes because we want
52   to read the velocity data it.
53   """
54   t_start = 0.0
55   t_stop = 3.0
56   delta_t = 1 / 1500
57   n_steps = int((t_stop - t_start) / delta_t)  # N steps
58
59   """
60   These constants are defined as they are
61   to be used in the weak form of the PDE.
62   """
63   dt_inv = fem.Constant(
64       domain=domain,
65       c=PETSc.ScalarType(1 / delta_t))
66   Rate = fem.Constant(
67       domain=domain,
68       c=PETSc.ScalarType(0.9))
69   D_coeff = fem.Constant(
70       domain=domain,
71       c=PETSc.ScalarType(0.03))
72
73   """
74   Here, we define the element type
75   and the function space to solve
76   for the concentration of the
77   reactant.
78   """
79   P1 = ufl.FiniteElement(
80       family="Lagrange",
81       cell=domain.ufl_cell(),
82       degree=1)
83   V = fem.FunctionSpace(
84       mesh=domain,
85       element=P1)
86   v1 = ufl.TestFunction(function_space=V)
87   u1 = fem.Function(V=V)
88   u1n = fem.Function(V=V)
89   u1.name = "concentration"
```

```
90   u1n.name = "concentration"
91
92   """
93   Here, we define a vector element
94   and a corresponding function space,
95   function for loading the
96   velocity (a vector) data from the
97   NS simulation.
98   """
99   Pvec = ufl.VectorElement(
100      family="Lagrange",
101      cell=domain.ufl_cell(),
102      degree=2)
103  W = fem.FunctionSpace(
104      mesh=domain,
105      element=Pvec)
106  w = fem.Function(V=W)
107
108
109  class SourceExpression:
110      """
111      We define the source location and its
112      magnitude that would go into the weak
113      form. The (x, y) location is specified
114      using 'ptSrc_xy' argument.
115      The magnitude is specified as 10.0.
116      """
117      def __init__(self, t_, ptSrc_xy):
118          self.t = t_
119          self.ptSrc_xy = ptSrc_xy
120
121      def eval(self, x):
122          values = np.full(x.shape[1], 0.0)
123          idx = ((x[0] - self.ptSrc_xy[0]) ** 2) + \
124                ((x[1] - self.ptSrc_xy[1]) ** 2) < 0.05 ** 2
125          values[idx] = 100.0  # Strength of the source
126          return values
127
128
129  # Inheriting from the source function class definition
130  source = SourceExpression(t_=0.0,
131                            ptSrc_xy=[0.25, 0.75])
132
133  # Interpolating into a function
134  f = fem.Function(V=V)
```

```python
135    f.interpolate(u=source.eval)
136
137    """
138    Here, we define the weak formulation
139    of the partial differential equation.
140    'w' is a function defined on a vector
141    function space 'W' that holds the value
142    of the velocity field. This velocity
143    is taken from a previous Navier-Stokes
144    simulation.
145    """
146    F = ((u1 - u1n) * dt_inv) * v1 * dx
147    F = F + dot(w, grad(u1)) * v1 * dx
148    F = F + D_coeff * dot(grad(u1), grad(v1)) * dx
149    F = F + Rate * u1 * v1 * dx
150    F = F - f * v1 * dx
151
152    """
153    The current advection reaction equation,
154    although solvable by a linear solver, we
155    deliberately do away with a nonlinear
156    solver because one may wish to model a
157    case where the rate of decomposition of
158    the contaminant may at times be proportional
159    to a nonlinear expression of the concentrations
160    of the reactants. Newton solver is good with them.
161    """
162    problem = fem.petsc.NonlinearProblem(
163        F=F,
164        u=u1)
165    solver = nls.petsc.NewtonSolver(
166        comm=MPI.COMM_WORLD,
167        problem=problem)
168    solver.rtol = 1e-6
169    solver.report = True
170
171    """
172    We use the open file format '.xdmf' to store
173    the results of the simulation. We first write
174    the mesh followed by the function. The mesh is
175    written just once; in the time loop we only
176    write the function values as the mesh gets
177    shared during visualization.
178    """
```

```
179  xdmfu = XDMFFile(
180      comm=domain.comm,
181      filename=resultPath + simulationName + ".xdmf",
182      file_mode="w",
183      encoding=XDMFFile.Encoding.HDF5)
184  xdmfu.write_mesh(mesh=domain)
185  xdmfu.write_function(
186      u=u1,
187      t=t_start)
188
189  """
190  We load the velocity data file from the Navier-Stokes
191  simulation. The velocity vector (u_i, u_j) is used
192  to realize the advection part of the equation.
193  Because the data was stored in HDF5 format, we employ
194  h5py library to load the data. This is done just before
195  'solving' the equation.
196  """
197  with h5py.File(
198          name=velocityDataPath, mode='r') as fr:
199      print("Name of the velocity variable is {}".format(
200          list(fr.keys())))
201
202      # Time stepping and solving
203      progress = tqdm.autonotebook.tqdm(
204          desc="Solving nonlinear system",
205          total=n_steps
206      )
207      t = t_start
208      for n in range(n_steps):
209          print("Doing {}/{} step".format(n, n_steps))
210          t = t + delta_t
211          u_data = fr['u'][n]
212          w.vector.array = u_data
213          r = solver.solve(u=u1)
214          u1n.x.array[:] = u1.x.array
215          xdmfu.write_function(
216              u=u1,
217              t=t)
218      xdmfu.close()
219  fr.close()
220  print("Done!")
```

From lines 35–46, we load the mesh that was generated and saved using the GMSH mesh-generating software. Lines 40–44 load the mesh object, the cell tags, and the facet tag objects. In our mesh we do not have any cell tags; we only have the facet tags and are required to identify the boundaries for setting up the boundary conditions. The dimension of the facets has to be 1 less than the dimension of the mesh, hence lines 45 and 46.

In lines 48–57, we define the time discretization parameters like the initial, and final and the steps at which to compute the solution. We specify to do 1500 calculations in the time interval of 0 to 3 s.

Next, in lines 59–71, we define the constants to be used later in the program. Here, "fem" is an alias for the "dolfinx" library in which the keyword "Constant" defines a constant for the loaded "domain" or mesh; "c" specifies the value of the constant and has a type "PETSc.ScalarType". This data type is native to the PETSc library that is internally used by "dolfinx" to solve the nonlinear problem.

In lines 73–90, we define the finite element types and family for the loaded mesh. The loaded mesh already contains "triangular" elements and is suitably referred to by the "domain.ufl_cell". The Lagrange basis shape functions of degree 1 have been specified—suitable for working with scalar-type data of the concentration variable. The object "V" is used to store the "FunctionSpace" object where the P1 elements of type "Lagrange" are defined; Function spaces, in finite element models, are a way to associate the cells of the mesh with the element type (Lagrange here). "v1, u1, u1n" are a test function, function, and function defined on the "V" function space. We also give a suitable name—"concentration" to the functions.

In lines 93–106, we define the finite element types and family for the loaded mesh. The loaded mesh already contains "triangular" elements and is suitably referred to by the "domain.ufl_cell". The Lagrange basis shape functions of degree 2 have been specified; a higher degree implies higher accuracy at the cost of extra computation. "W" and "w" define the "Function Space" and a "Function" on the domain. "V" is an argument here which expects a "FunctionSpace" object. "Pvec" has been defined to be a vector because we intend to load the pre-saved velocity (vector) data into it from the Navier-Stokes simulation.

In lines 109–126, we define a class for the source function. A class facilitates the definition of a time-dependent source. The "__init__" method is used to assign an initial value of the time and coordinates of the source location. The "__eval__" method is used to evaluate the value of the source. Lines 123–124 mark the coordinates as Boolean "True" if the radius is less than 0.05, helping to specify a larger circular source. If indices where the source is found to be true, a value of 100 is assigned. In lines 129–135, we invoke the source class with an initial time value of 0.0 and location coordinate (0.25, 0.75). In lines 134–135, we define a function on the "V" function space and use the interpolate method to assign the value 100 at designated (index = True) coordinates.

In Lines 137–150, we define the variational formulation for the advection, diffusion, reaction problem.

In lines 152–169, we set up the solver object; lines 162–164 specify the problem type, "NonlinearProblem"; the variational form, F and the dependent variable we wish to solve for, $u1$.

Lines 172–187 define the file output operations by specifying the "w" option—the write operation; "domain.comm" has to do with parallel processing (currently set for serial operation). Encoding specifies the file format for storing the field variable data (here, binary hierarchical data format or HDF5). We first write the mesh object followed by a value of the function and the associated time step in lines 185–187.

From lines 189–220, we specify the time loop for solving the problem sequentially. In lines 197–200, we use the "with" context manager to read (r)the velocity data file in the "fr" object. Lines 202–206 are used to display the progress bar. In line we increment the time with a value = "delta_t" because the function corresponding to "t_inital" is already written in line 187. In line 211, we extract the value at the nth step from a dictionary-type object with an "u" key; this value is assigned to the "w" function in line 212. Once the velocity vector is loaded, in line 213, we invoke the "solve" method to solve for "u1"; the solution of which is, next, stored into "u1n". This "u1n" then serves to provide the concentration value for the next time step in the loop.

At last, we close the file objects to save the data into the hard disk.

14.3.3   Post-processing

All the post-processing has been done in the Paraview software. The program in Sect. 14.3.2 saves the results in ".xdmf" format. These files are open-source file formats that can be readily loaded by Paraview. The file contains simulated data for each time step from $t = 0.00$ to $t = 3.00$ in 1500 time steps. This has been kept the same as the Navier-Stokes simulation. Figure 14.3 has been plotted by selecting the "Surface"-type representation and "Coloring" by the variable's magnitude in the loaded mesh. Figure 14.3a–f was generated by selecting appropriate time values from the "Time" dropdown menu in the software. The Paraview application uses "Cool to Warm" colormap by default. A better alternative is the "Turbo" colormap. One can see the spreading of the contaminant even at end-time values.

Figure 14.4 was plotted by selecting the "Surface With Edges" representation and setting the "Coloring" option to "Solid Color". The pink-colored nodes are produced by doing a "Select Points Through (g)" operation in the rendered window for the mesh. These nodes represent the top-left bank of the aquifer and would be used to extract the concentration values for all time steps. Data at each of these nodes is thus a time series enabling further computation. Figure 14.5 is produced by doing the following statistical computations—first and third quartile, mean, maximum, and minimum. A separate Python program was used to generate the figures.

14.3.4   Discussion of the Results

In this section, we have demonstrated a contaminant transport model where the contaminant undergoes advection, diffusion, and reaction. The result has been shown in Fig. 14.3a–f. In each a contour plot of the concentration u has been shown in a 2D plane. The source of the contaminant has been placed at the location (0.25, 0.75) (see lines 130 and 131 in code Sect. 14.3.2). The program simulates the transport of the contaminant to other parts of the domain starting from this source location. It can be

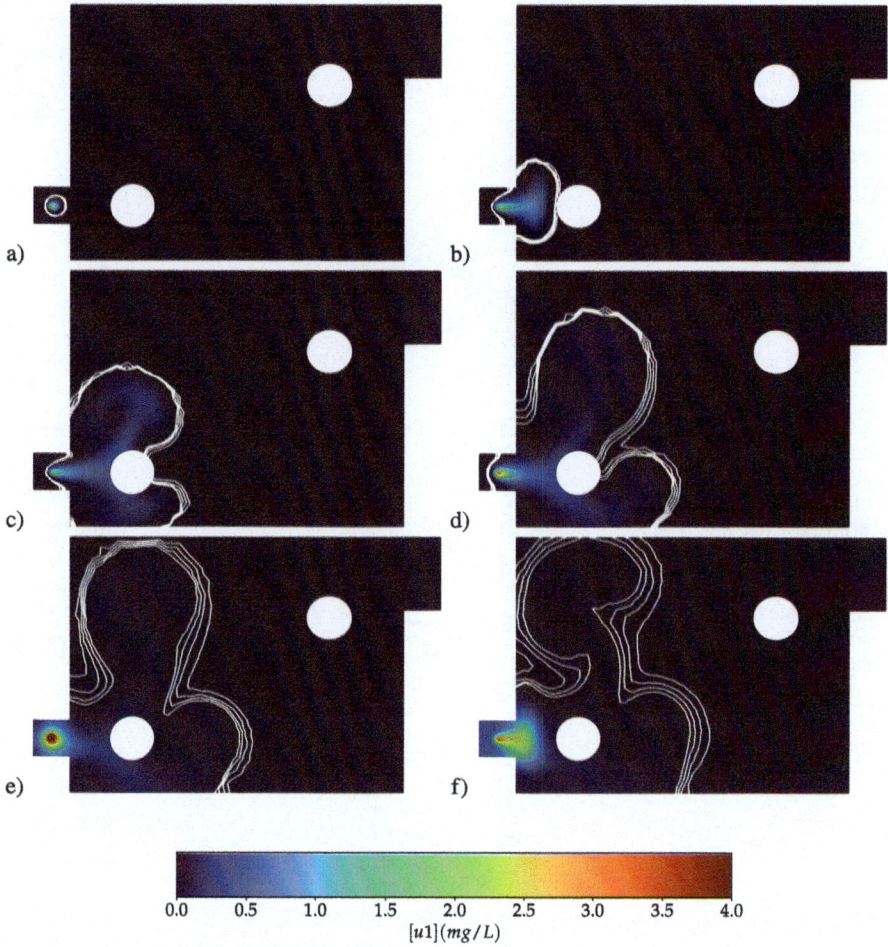

Fig. 14.3 Concentration field as computed by Program Sect. 14.3.2 at 6 different times, $t = 0.03$ s, 0.33 s, 1.0 s, 1.66 s, 2.33 s, 3.00 s. One can notice specific patterns that emerge due to the presence of the two circular-shaped islands imposing the no-slip boundary condition along with the surrounding boundaries

seen that with time the contours capture a larger area where their shape is governed by the velocity vector. This velocity contributes to the advection component of the combined transport process.

In addition to the location of the source the magnitude of the source has also been specified (see line 125 in code Sect. 14.3.2). This is responsible for the amount of the contaminant being added at the location. A higher value ensures an increased addition of the contaminant and vice versa. Likewise, multiple sources with varying strengths can be specified in the domain.

Fig. 14.4 We intend to record the concentration statistics on the top-left bank of the domain over time. The figure shows the selected nodes in pink used to do the calculations

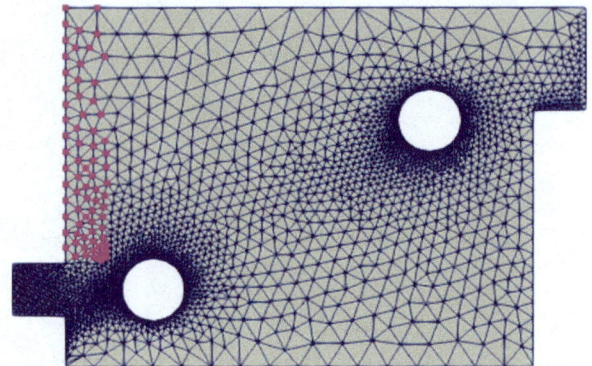

Fig. 14.5 Concentration data statistics for three contaminant transport models: **a** diffusion reaction, **b** diffusion-advection, and **c** advection-diffusion reaction. The inside color bands (dark gray) indicate the 1st and 3rd quartiles. The outer color bands (light gray) indicate the minimum and maximum of the concentration recorded at the nodes. The selected nodes are shown in Fig. 14.4

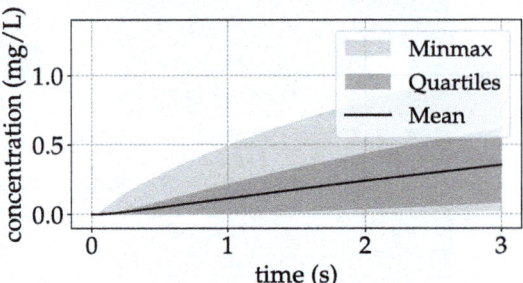

a)

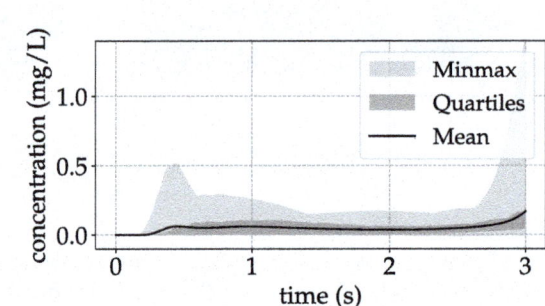

b)

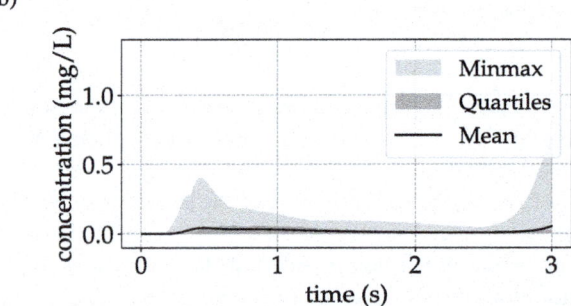

c)

The patterns of the contours appear to be quite interesting when compared to the ones obtained using previous models of Sects. 14.1 and 14.2. The constants D and R play a significant role in determining the amount of diffusion and the rate at which the contaminant decays. A higher value of D would ensure a diffusion-dominated transport process. The rate of reaction R helps specify how fast the contaminant decays with a given concentration; a smaller value would induce an advection, diffusion-type pattern whereas a higher value would diminish the effect of those processes.

Interestingly enough the geometrical pattern of the contours appears similar to that of the advection, diffusion process. The reason for this is quite obvious—the reaction rate (currently 0.9) is not strong enough to mask the contribution of advection and diffusion processes.

It is often of interest to monitor the evolution of the concentration of the contaminant as a function of time. With the developed models, we perform this task for the top-left bank of our aquifer model. We track the evolution of the concentration at the computational nodes as shown in Fig. 14.4. In Fig. 14.5, we show the statistics of the concentration values recorded due to the three models.

In Fig. 14.5a, the diffusion, reaction model shows a continuous increase in the concentration value. The mean value bears a positive slope and rises quickly to a value of around 0.3. Parts (b) and (c) show a similar pattern in their statistics due to a moderate value of the reaction rate, although one can still notice that the magnitudes in the advection, reaction model in part (b), the statistics are clearly higher than generalized advection, diffusion, reaction model in part (c).

Chapter 15
Conclusion

Python has emerged as one of the most popular languages for applications in hydrology, environment, and climate. In this textbook, we have provided extensive codes that can be used for common data analysis and numerical modeling needs.

Part I started with the basics of Python. We highlighted the advantages of open-source programming languages like Python and how they cater to the needs of various data types. The importance of virtual environments was stressed upon while emphasizing the need for integrated programming environments like the Jupyter notebook, the Anaconda Python distribution, and its associated package manager. The basic syntax for programming was covered highlighting Python's functional capabilities and its readiness to handle multiple data structures. We also spoke about the data manipulation capabilities offered by the versatile Python library—Pandas along with an illustration of the powerful plotting library—Matplotlib.

Part II is concerned with statistical modeling in hydrology. The implementation of several programs for statistical data analysis proves the versatility of the scientific computation packages in Python. We used libraries such as the SciPy, statsmodels, and scikit-learn for performing curve-fitting, doing regression analysis, and fitting time series models. A chapter was dedicated to hypothesis testing showcasing the simplicity with which Python language supports statistical testing. The uncertainty estimation chapter showcased the flexibility of Python libraries like SciPy in quantifying the uncertainty of the data. This was illustrated through codes that could output confidence intervals, prediction intervals, and Monte Carlo uncertainty propagation.

The strength of Python in numerical modeling was highlighted in Part III of the book where we leveraged the simplicity of the FEniCSx to specify a finite element problem in different scenarios. The seepage flow simulation and the groundwater flow simulations were carried out. We demonstrated how using the FEniCSx finite element library simplified the formulation of the numerical problem. We also showed how the Numpy library helped in performing the channel flow simulations and solving 2D shallow water equations. Visualization libraries like PyVista and Matplotlib helped plot the results and create 3D visualizations at different stages.

© The Author(s), under exclusive license to Springer Nature Singapore Pte Ltd. 2024   287
A. Kumar and M. Saharia, *Python for Water and Environment*, Innovations in Sustainable Technologies and Computing, https://doi.org/10.1007/978-981-99-9408-3_15

In Part IV, the FEniCSx proved its versatility in solving contaminant transport problems. We delved deep into formulating 2D transport problems involving advection, diffusion, and reaction and their combinations. The versatility of another open-source package GMSH, was also evident while creating the computational domains for various simulations. The package could create error-free computational meshes ready to be used by a FEM program. The result was an end-to-end FEM design, processing, and post-processing framework that can well be adapted to different numerical modeling problems.

The blend of in-depth documentation and open-source nature makes Python an unparalleled choice for dealing with complex challenges in water and environment science. As the focus on sustainable water and environmental management sharpens globally, this book will equip professionals with the tools and knowledge to be at the forefront.

The manufacturer's authorised representative in the EU is Springer
Nature Customer Service Centre GmbH, Europaplatz 3, 69115 Heidelberg,
Germany. If you have any concerns regarding our products, please
contact ProductSafety@springernature.com

Printed and bound by CPI Group (UK) Ltd, Croydon, CR0 4YY
27/04/2026
02097580-0005